CURRICULUM AND EVALUATION

S T A N D A R D S

FOR SCHOOL MATHEMATICS

ADDENDA SERIES, GRADES 5–8

DEALING WITH DATA AND CHANCE

Judith S. Zawojewski

with

Gary Brooks
Lynn Dinkelkamp
Eunice D. Goldberg
Howard Goldberg
Arthur Hyde
Tess Jackson
Marsha Landau
Hope Martin
Jeri Nowakowski
Sandy Paull
Albert P. Shulte
Philip Wagreich
Barbara Wilmot

Consultants

Diana Lambdin Kroll
Frank K. Lester, Jr.
Kenneth P. Goldberg

Frances R. Curcio, Series Editor

1906 Association Drive, Reston, VA 20191-9988
(703) 620-9840; (800) 235-7566; www.nctm.org

Seventh printing 2003

Library of Congress Cataloging-in-Publication Data:

Zawojewski, Judith S.
Dealing with data and chance / Judith S. Zawojewski ; with Gary Brooks...[et al.]
p. cm. — (Curriculum and evaluation standards for school mathematics addenda series. grades 5–8)
Includes bibliographical references.
ISBN 0-87353-321-6
1. Probabilities—Study and teaching (Elementary) I. Brooks, Gary D. II. Title. III. Series.
QA273.2.Z39 1991
372.7—dc20 91-5036
CIP

The publications of the National Council of Teachers of Mathematics present a variety of viewpoints. The views expressed or implied in this publication, unless otherwise noted, should not be interpreted as official positions of the Council.

Printed in the United States of America

TABLE OF CONTENTS

FOREWORD

In March 1989, the National Council of Teachers of Mathematics officially released the *Curriculum and Evaluation Standards for School Mathematics* (NCTM 1989). The document provides a vision and a framework for strengthening the mathematics curriculum in kindergarten to grade 12 in North American schools. Evaluation, an integral part of planning for and implementing instruction, is an important feature of the document. Also presented is a contrast between traditional rote methods of teaching, which have proven to be unsuccessful, and recommendations for improving instruction supported by current educational research.

As the *Curriculum and Evaluation Standards* was being developed, it became apparent that a plethora of examples would be needed to illustrate how the vision could be realistically implemented in the K–12 classroom. A Task Force on the Addenda to the *Curriculum and Evaluation Standards for School Mathematics*, chaired by Thomas Rowan and composed of Joan Duea, Christian Hirsch, Marie Jernigan, and Richard Lodholz, was appointed by Shirley Frye, then NCTM president, in the spring of 1988. The Task Force's recommendations concerning the scope and nature of the supporting publications were submitted in the fall of 1988 to the Educational Materials Committee, which subsequently framed the Addenda Project for NCTM Board approval.

Following the release of the *Curriculum and Evaluation Standards* and on the basis of the recommendations of the NCTM Task Force (Rowan 1988), writing teams were assigned to develop addenda to provide teachers with classroom ideas for translating the standards into classroom practice. Three writing teams were formed to prepare materials for grades K–6 (Miriam A. Leiva, editor), grades 5–8 (Frances R. Curcio, editor), and grades 9–12 (Christian R. Hirsch, editor).

The themes of problem solving, reasoning, communication, connections, technology, and evaluation have been woven throughout the materials. The writing teams included K–12 classroom teachers, supervisors, and university mathematics educators. The materials have been field tested in an effort to make them "teacher friendly."

Implementing the *Curriculum and Evaluation Standards* will take time. The Addenda Series was written to help teachers as they undertake this task. Furthermore, the Addenda Series is appropriate for use in the in-service staff development as well as for use in preservice courses in teacher education programs.

On behalf of the National Council of Teachers of Mathematics, I would like to thank all the authors, consultants, and editors who gave willingly of their time, effort, and expertise in developing these exemplary materials. In particular, I would like to acknowledge gratefully the work of Judith Zawojewski and her contributing authors, Gary Brooks, Lynn Dinkelkamp, Eunice Goldberg, Howard Goldberg, Arthur Hyde, Tess Jackson, Marsha Landau, Hope Martin, Jeri Nowakowski, Sandy Paull, Albert Shulte, Philip Wagreich, and Barbara Wilmot, for preparing the *Dealing with Data and Chance* manuscript. Gratitude is also extended to the individuals who reviewed the manuscript during the stages of development: Cheryl Brothers, Thomas Cook, Larry Elchuck, Mary M. Lindquist, Faye Tatel, and M. B. Ulmer. Finally, this project would not have materialized without the outstanding technical support supplied by Cynthia Rosso and the NCTM staff.

Bonnie H. Litwiller, *Addenda Project Coordinator*

PREFACE

The purpose of *Dealing with Data and Chance*, as well as of the other books in the Grades 5–8 Addenda Series, is to provide teachers with ideas and materials to support the implementation of the *Curriculum and Evaluation Standards for School Mathematics* (NCTM 1989). In addition to this book, the other four publications in this series are *Understanding Rational Numbers and Proportions* (Bezuk, forthcoming); *Geometry in the Middle Grades* (Geddes, forthcoming); *Patterns and Functions* (Phillips 1991); *Developing Number Sense in the Middle Grades* (Reys 1991). These books are *not* an outline of a middle school curriculum, but rather, they present several topics and activities to exemplify the ideas advocated in the *Curriculum and Evaluation Standards*, and they provide examples to help students make the transition from elementary to high school mathematics.

The unifying themes of the *Curriculum and Evaluation Standards,* the characteristics of a new classroom learning environment, and the role of evaluation are described below. The discussion in each of these parts will refer to examples in this book.

Unifying Themes

The unifying themes of the *Curriculum and Evaluation Standards* include mathematics as problem solving, mathematics as communication, mathematical reasoning, and mathematical connections. These themes are not separate, isolated entities, but rather, they are all interrelated. The examples and illustrations presented in this book are designed to demonstrate the interrelationships by weaving these themes throughout the activities as well as by providing ideas for incorporating technology and evaluation techniques.

Mathematics as problem solving. Although problem solving has been a goal of mathematics instruction throughout the years, it became the focus of attention with the advent of NCTM's *An Agenda for Action* (NCTM 1980). The *Curriculum and Evaluation Standards* reaffirms the importance of problem solving in mathematics instruction.

The excitement of learning and applying mathematics is generated when problems develop within the context of a situation familiar to students. Allowing them to formulate problems as they naturally arise within the context of everyday experiences gives them the opportunity to put mathematics to work, observing its usefulness and its applicability (e.g., Illustration 1). However, "not all problems require a real-world setting. Indeed, middle school students often are intrigued by story settings or those arising from mathematics itself" (NCTM 1989, p. 77). For example, see Illustration 3.

Mathematics as communication. Although the formal language of mathematics is concise, high in concept density, and may be seemingly foreign, students should have the opportunity to bring meaning to mathematics on the basis of their experiences. Allowing them to talk about their experiences and how they relate to mathematics concepts, listen to each other as they share ideas, read mathematics in various formats (e.g., number sentences, graphs, charts), and write about mathematical situations affords students the opportunity to compare experiences, clarify their thinking, and develop an understanding of how the mathematics they study in school is related to the mathematics they experience in the "real world". This requires the integration of the four language arts—

speaking, listening, reading, and writing—with the mathematics lesson (e.g., Illustration 5).

Communicating in mathematics requires a common language and a familiarity with modes of representing mathematical ideas. Depending on students' "comfort level," oral language, prose, manipulatives, pictures, diagrams, charts, graphs, or symbols can be used to communicate ideas (e.g., Illustration 6). They should be encouraged to translate from one mode to another.

Mathematics as reasoning. During the middle school years, students should have opportunities to develop and employ their abilities in logical and spatial reasoning, as well as in proportional and graphical reasoning. The development of a student's reasoning ability occurs over a period of time. We can observe extreme differences between students in grade 5 and students in grade 8. As a result, instructional approaches must reflect these differences.

Depending on children's readiness, exploratory activities, experiments, and projects may require them to give a descriptive account of what they observe, an informal argument based on empirical results, or a formal proof supporting a conjecture (Hirsch and Lappan 1989). Active learners should be constantly involved in questioning, examining, conjecturing, and experimenting (e.g., Illustrations 9 and 10).

Mathematical connections. Traditionally, mathematics has often been presented as an isolated set of rules to be memorized. The *Curriculum and Evaluation Standards* suggests that mathematics be presented as an integrated whole. Students should observe the interrelatedness among branches of mathematics: number theory, geometry, algebra, probability, and so on (e.g., Illustration 12). Furthermore, students should become aware of how mathematics is related to other disciplines such as science, art, literature, music, social studies, business, and industrial technology (e.g., Illustrations 2 and 13).

There is no guarantee that allowing students to explore, create, and experiment within the context of a problem-solving setting will lead them to discover connections between and among mathematics concepts. Teachers may need to guide students in discovering connections and to elicit these connections explicitly. Students who recognize connections within mathematics and with other disciplines can understand and appreciate the logical unity and the power of mathematics (Steen 1989).

A New Classroom Environment

Implementing the *Curriculum and Evaluation Standards* requires a new way of teaching. The traditional teacher roles of authority figure and information disseminator must change to learning facilitator and instructional decision maker.

Knowledge about students and how they learn mathematics can contribute to establishing a conducive learning environment for middle school students. The teacher selects the instructional objectives based on knowledge of his or her students, knowledge of mathematics, and knowledge of pedagogy (NCTM 1991). After selecting the instructional objectives, the teacher must decide how to deliver the content. Is the use of manipulatives appropriate? Is the use of technology appropriate? Is a cooperative learning setting appropriate?

Appropriate use of manipulatives. Manipulatives are multisensory tools for learning that provide students with a means of communicating ideas by allowing them to model or represent their ideas concretely. Using manipulatives, however, does not guarantee the understanding of a mathematics concept (Baroody 1989). After allowing students to explore using manipulatives, teachers must formulate questions to elicit the important, "big" mathematical ideas that enable students to make connections between the mathematics and the manipulatives used to represent the concept(s) (e.g., Illustration 4).

Appropriate use of technology. Developments in technology have made the traditional, computation-dominated mathematics curriculum obsolete. As a result, technology has been given a prominent place in the *Curriculum and Evaluation Standards*, in terms of which technology should be made available for use in the classroom and how technology should be used in mathematics instruction. It is expected that at the middle school level, students will have access to appropriate calculators, and computers should be available for demonstration purposes as well as for individual and group work (NCTM 1989, p. 8). It should be noted that new advances in technology are being made constantly. Teachers should keep abreast of new developments that support mathematics instruction.

In this book, suggestions are made for integrating the use of calculators to facilitate and expedite computation (e.g., Illustration 11), and to explore and discover the meaning of mathematics concepts (e.g., Illustration 8). Specifically, simple four-function calculators and scientific calculators are recommended for use in many activities.

Recommendations for integrating computers are also presented. In particular, suggestions are made for using the computer to simulate experiments (e.g., Illustration 4) and for using software to develop survey questions (e.g., Illustrations 1 and 5), to graph and analyze data (e.g., Illustration 6), and to manipulate data on a spreadsheet (e.g., Illustrations 7 and 12). Also, depending on their abilities and background experiences, students may be assigned to write a program to generate random numbers (e.g., Illustration 1).

Appropriate use of cooperative learning groups. Traditionally, mathematics has been taught as a "solo," isolated activity, yet in business and industry, mathematicians often work in teams to solve problems and attain common objectives (Steen 1989). Allowing students to work in cooperative groups affords them the opportunity to develop social and communication skills while working with peers of various ethnic, religious, and racial groups.

Cooperative learning environments, characterized by students working together and interacting with each other, contribute to internalizing concepts by forcing the students to defend their views against challenges brought by their peers. The value of this approach is supported by the work of Vygotsky ([1934]1986) who discussed the increasingly interrelated nature of language and cognition as children grow.

Cooperative groups usually contain three to five students and may be established for various lengths of time (Artzt and Newman 1990). Unlike most traditional small group instruction in reading and mathematics, cooperative learning groups are heterogeneous and everyone must work together for the common good of all. Students who understand the concept being discussed are responsible for explaining it to those who do not understand. When using cooperative groups, teachers must consider new ways of evaluating performance to ensure the success of instructional objective(s).

The Role of Evaluation

Making changes in the content and methods of mathematics instruction will also require making changes in why and how students' work is evaluated. Evaluation should be an integral part of instruction and not be limited to grading and testing. There are at least four reasons for collecting evaluation information:

- to make decisions about the content and methods of mathematics instruction (e.g., Illustration 2, Activity 1);
- to make decisions about classroom climate (e.g., Illustration 1, Activity 1);
- to help in communicating what is important (e.g., Illustration 7, Activity 3);
- to assign grades (e.g., Illustration 1, Activity 5).

In other words, evaluation includes much more than marking right and wrong answers. It "must be more than testing; it must be a continuous, dynamic, and often informal process" (NCTM 1989, p. 203). What methods can be used for evaluation purposes? The *Curriculum and Evaluation Standards* recommends that teachers use a variety of types of evaluation. Examples can be found throughout the text as indicated: (1) *observing and questioning students* (e.g., Illustration 9, Activity 3); (2) *using assessment data reported by students* (e.g., Illustration 12, Activity 1); (3) *assessing students' written mathematics work* (e.g., Illustration 1, Activity 5); and (4) *using multiple-choice or short-answer items* (e.g., Illustration 6, Activity 4). Teachers might, for example, observe and question students to assess their understanding and to gain insight into their feelings and their beliefs about mathematics; use holistic scoring techniques for a focused assessment of students' written problem-solving work; or collect information through students' responses to short-answer questionnaires or through written assignments such as journal entries or brief essays. Use of these multiple methods of collecting assessment data will contribute to a thorough evaluation of students' work. These and other evaluation techniques are discussed in more detail in Lester and Kroll (1991).

This brief description of the unifying themes of the *Curriculum and Evaluation Standards*, the characteristics of a new learning environment, and the role of evaluation is provided as a starting point to understand and to appreciate the ideas that are presented in this book. It is hoped that these ideas provide a foundation for developing concepts in data analysis and chance in the middle grades and generate interest among teachers for improving instructional and evaluation techniques in mathematics.

Frances R. Curcio, Editor
Grades 5–8 Addenda Series

INTRODUCTION

The role of data and chance in school mathematics is rapidly changing both in increased emphasis and in methods of instruction. *Dealing with Data and Chance* first considers how people naturally use their understanding of data and chance in their daily experiences and then addresses the role of data and chance in middle school mathematics. The major section of the book, chapters 1 through 5, makes a case for building on these natural abilities for learning from explorations with data and chance. Five themes are illustrated with classroom activities: data gathering by students (Illustrations 1, 2, 3, and 4); communication (Illustrations 5 and 6); problem solving (Illustrations 7 and 8); reasoning (Illustrations 9, 10, and 11); and connections (Illustrations 12 and 13). Each illustration contains several activities and extensions that require one or more class periods. Blackline masters, available for many of the activities, are found in Appendix 1.

Natural Learning Experiences with Data and Chance

Throughout life, people make decisions based on their own informal collection of data. For example, consumers often base their choice of brands of food and makes of cars on an informal analysis of past experiences and on information gathered from acquaintances. Even a baby collects and uses personal data in learning. For example, a baby collects data by repeatedly dropping a spoon from her high chair. The baby soon learns that the spoon will *certainly* fall to the floor every time she drops it. *Uncertainty* is also learned by collecting personal data. For example, the baby finds that *usually* somebody nearby will pick up and return the spoon to her. Although the baby has learned that the return of the spoon is an uncertain event, she has also found a way to increase the chances that the spoon will be returned by crying and reaching for the spoon until someone does indeed return it. The baby, in essence, has gathered data and learned about certain and uncertain events from data.

Later in life, a child learns to play games and develops a sense of "fairness," which is related to the notion of equally likely events. Three-year-olds believe that "fair" means that they win. By first grade, children believe that the toss of a coin is a fair way to decide who gets the first choice for a piece of cake. The child's informal method of collecting and analyzing data leads to effective evaluations and judgments about the fairness of a game and the likelihood of winning or losing under specific conditions.

Adults generally make decisions based on their perceived likelihood of an event. One way these perceptions are formed is by informally gathering and analyzing data for a specific situation. Consider Alice, who has three convenient routes to her workplace. After a year of trials and experiences, Alice has settled on the route that she uses most often. When asked why she takes that route to work every day, she answers that she had found this to be the fastest way during rush hour. Her decision is based on the informal collection and analysis of data from past trial runs of various routes and on a recognition that travel time can vary.

Like Alice, people are generally capable of making good decisions on the basis of effective predictions of chance using relatively small sets of data. However, there is evidence that over time we develop a small number of limited strategies that we apply automatically and repeatedly in varied contexts and to situations involving larger amounts of data. Often the application of the strategy is ineffective or totally erroneous, and we do not even recognize that we have not maximized our chances of obtaining the optimal result. For example, people are more swayed by personal anecdotes than by reports from large samples of data. Kahneman, Slovic, and Tversky (1982) have identified and investigated a number of decision-making methods, which for the most part produce satisfactory results, but which are actually based on faulty reasoning. Interestingly, they found that even people who were professionally well versed in the study of statistics used these same erroneous procedures when outside their professional context.

As an illustration appropriate for middle grade students, consider the process that Tom and Rita used in selecting a new car. Wanting to maximize their chances of buying a good, reliable car, they first went to the library to find out the makes and models recommended by a well-respected source, *Consumer Reports*. The report made recommendations based on a sample of hundreds of car owners who were surveyed about comfort, maintenance costs, and so on. Then, Tom and Rita sought the advice of a few friends and family members. They found that three people they knew favored a car not highly recommended by *Consumer Reports*. The others recommended different makes. After carefully considering what everyone had said, Tom and Rita decided to go with the recommendations of the people they *knew*.

This decision-making process may seem perfectly reasonable at first glance, but the process is actually faulty. *Consumer Reports* and Tom and Rita asked similar questions in their "surveys." The main difference in approach was that the sample used in *Consumer Reports* was much larger and more diverse than the sample of Tom and Rita's friends and family. Their decision to follow the advice of people they knew did not maximize their chances of getting the "best buy" because the information from *Consumer Reports* would actually be the best predictor of selecting a "good" car. Tom and Rita were likely swayed by the richness of the descriptions made by the people they knew and trusted. Nisbett and Ross (1980) have documented that adults are generally influenced more by the "vividness" of anecdotes than by data taken from large, diverse samples.

These accounts of daily encounters with data bring two instructional implications to mind. First, classroom experiences should build on students' natural abilities to use data to solve problems in everyday situations of uncertainty. Second, sound reasoning and decision-making experiences with large sets of data should be included in instruction. Students need guidance in learning to interpret and analyze large sets of data, which may seem less vivid or real than smaller sets of anecdotal data. When data have been gathered, organized, represented, and analyzed by someone else, students need to rely on their own experiences to question data gathering processes, sampling biases, conclusions, and so on. Through their own experiences, students build critical thinking skills that help them make decisions in uncertain situations.

The Role of Data and Chance in the Middle School

In the society for which we are educating children, the availability of information is exploding—computers are able to store, sort, and analyze massive amounts of data. Virtually all businesses, manufacturing firms, services, and professions require decision-making capabilities under uncertain conditions. The abilities to analyze and use data to maximize chances for success are critical problem-solving tools that are becoming increasingly important. The implications for school mathematics is reflected in the *Curriculum and Evaluation Standards for School Mathematics* (NCTM 1989), in both the increased emphasis in these topics and the guidelines for modifying methods of instruction.

The *Curriculum and Evaluation Standards* advocates teaching probability and statistics in a holistic manner. Students should learn relevant concepts and skills in the context of applications and with connections to other mathematics and nonmathematics topics. Standards 10 and 11 for grades 5 through 8 (NCTM 1989, pp. 105–9) illustrate the shift from direct teaching of isolated techniques to student-centered experiences, explorations, and applications.

Implementing the Standards

The *Curriculum and Evaluation Standards* not only calls for an increased emphasis on curricular topics dealing with data and chance but also emphasizes a shift in the methods of teaching the related concepts and skills. Burrill's (1990) guidelines for implementing the *Curriculum and Evaluation Standards* include suggestions that reflect that shift in instructional practices and furnish a foundation for the activities in this book.

The major part of this book is divided into five sections, each reflecting an important emphasis for exploring data and chance in the middle school and each including illustrations of appropriate classroom activities. The first section focuses on students gathering their own data through surveys, experiments, and simulations. The next four sections focus on the themes of the *Curriculum and Evaluation Standards*: communication, problem solving, reasoning, and connections. Although each section focuses on a particular feature of the *Curriculum and Evaluation Standards*, every illustration incorporates multiple themes.

CHAPTER 1
DATA GATHERING BY STUDENTS

When students gather their own data, they become intimately familiar with the information. They are then able to move from individual experiences with vast amounts of detail to an experience in which they tally, represent, and combine those individual pieces of information. During the process, they develop an understanding of what it really means to organize and summarize data. The implications of grouping and combining data for the sake of gaining a larger, more generalized picture take on meaning. When students draw conclusions based on their experience, they develop the skills needed to become critical consumers of someone else's data summaries and analyses. All citizens are bombarded daily with results from studies, surveys, and polls that someone else has done. Only through personal experiences with the whole process can students develop the critical thinking skills to question, analyze, and interpret data from outside sources.

Large data collection projects readily embed all four themes of the *Curriculum and Evaluation Standards*. Communication plays a major role in formulating questions and verbally analyzing data. Further, the charts and graphs that organize and represent data are forms of communication for users of information. Problem solving permeates the whole process as students decide on interesting topics, formulate questions, plan the collection of data, implement plans, analyze results, make conjectures about their original questions, and decide how to formulate answers. Statistical and probabilistic reasoning are used as students conjecture about cause-and-effect relationships, search for alternative explanations for data trends, and make decisions under uncertain conditions. Finally, large data collection projects are inherently connected to a variety of academic disciplines, since the context of the question is independent of the process of gathering, organizing, summarizing, and analyzing data. Research projects provide students with active experiences in dealing with information and data firsthand. Such projects are easily implemented with all students at the classroom level and nationwide statistics competitions have become popular. Hawkins (1987) reported that nine- to nineteen-year-old students in the United Kingdom regularly address social, commercial, and scientific questions in the Annual Applied Statistics Competition. Recently, an American Statistical Prize Competition, modeled on the British version, was started in the United States (Cameron 1987).

Students can become involved in gathering and processing their own data through surveys, experiments, and simulations. Surveys are a nice way to begin the process with middle school students. The information is usually descriptive, requiring only that students find valid means of obtaining the desired data. Experiments are somewhat more sophisticated because students not only use descriptive techniques but also must design experiments using the scientific method. Simulations are similar to experiments but can be more sophisticated, since they use random number devices—number cubes, spinners, random number tables, and computer programs—to represent, or model, real-world situations.

SURVEYS

The spirit of the *Curriculum and Evaluation Standards* is captured by having students conduct their own surveys. Children's natural curiosity is

◆ ◆ ◆ ◆ ◆ ◆ ◆ ◆

piqued when they have information about topics of interest to them. Students enjoy the whole process of developing, conducting, and analyzing surveys. The following survey project was developed by a fifth-grade teacher as a long-term investigation implemented over six weeks.

ILLUSTRATION 1: SURVEYS, STATISTICS, AND STUDENTS: AN INTERDISCIPLINARY UNIT

Sandy Paull

This unit introduces students to the concept of average by looking for a profile of the "average" student in their school. During this project, students also see statistics used in the real world, standard curriculum skills used in application, and quantitative ideas used in other disciplines.

***CONNECTIONS:* Surveys are inherently interdisciplinary. The questions usually deal with nonmathematics topics, and the information is usually quantified for ease in organizing, summarizing, and analyzing.**

Materials

computer word-processing or survey-creating capabilities
centimeter grid paper
long lengths of string for each group of students
poems: "In the Middle" (McCord 1975, p. 7)
"Who's Who" (Viorst 1981, p. 14)

Preparation

Set up a word-processing file in which students can easily enter their survey question. Cut one seven-meter-long string for each group of students. Obtain the poems to read to the class.

***TECHNOLOGY:* Software has been designed for classroom use in developing survey questions, collecting data, and representing data.**

Activity 1: Introducing "Average"

1. Poems that give a delightful introduction to the concept of "average" are "In the Middle" by David McCord and "Who's Who" by Judith Viorst. Each poem refers to the idea of being average. Read the poems to the class. Ask the students to classify themselves as average or not. Lively discussions involving some value judgments are likely to occur as students share their ideas and explanations. Follow the discussion by assigning students to write a short essay, poem, or story about being average or describing the "average student" in their school.

2. *Evaluation.* Long-term projects furnish an opportunity for students to keep in a notebook a daily log describing what they have learned, the questions they still have, and their feelings about the activities and the assignments. Weekly reading of the students' logs can provide information to teachers for making decisions about further instruction. For example, some students in this illustration used *average* to reflect a social value: it is "good" or "not good" to be average. Because in mathematics the term is purely descriptive, it was helpful for the teacher to know that students were using the term for different purposes. The teacher was able to explore explicitly and purposely the subtle differences between the uses of *average*. Teachers should expose students to the idea that the meaning of terms can change subtly in different settings.

Activity 2: What Is a Survey?

1. Ask students to share their knowledge of polls and surveys. Topics commonly mentioned are TV ratings, predictions for presidential elections, and marketing surveys conducted in shopping malls. Ask students to share their personal, informal use of polls or surveys. For example, in deciding what kind of pizza to order for a recent birthday party, they may have asked all their friends which ingredients they liked or did not like.

2. Summarize by asking students to answer the following questions in writing:

What is a poll or a survey?

Why would someone want to conduct a survey?

What kind of questions do you think they ask?

What kind of information is obtained from the poll or the survey?

How do you think the information is used?

Activity 3: Developing the Project Goal

1. Share the project goal with the students: The class as a whole is to determine the profile of an "average student" in their school. Suggest that an effective way to obtain an honest, accurate picture is to do a survey. The survey is to be prepared and administered by students in the class. The first task is to figure out what information is needed.

2. Split the class into small groups to brainstorm the types of information needed. Groups should report their ideas, and the class as a whole should develop a master list of characteristics to investigate—height, favorite food, number of hours of TV watched per day, and so on.

Activity 4: Developing Class Survey Questions

PROBLEM SOLVING: As students develop their own questions for use in the surveys, they are involved in "problem formulation."

1. Ask the class as a whole to choose five characteristics from the list developed in Activity 3. Use student-suggested questions to form a five-question survey that all students should answer individually. Share the responses as a class and look for ways to tally and organize the responses. Note when responses are easy to tally (e.g., yes/no responses) and when they are difficult to combine or organize (e.g., open-ended responses). Decide why difficulties were encountered, and discuss ways to rewrite the items to make combining and tallying the data easier.

2. Administer the "new" five-question survey, but this time students should imagine answering the questions as though they were someone else. "Perspective-taking" can help students identify when questions may mean different things to different groups of respondents. For example, given the item "Do you like to read?," pose the following questions for discussion:

COMMUNICATION: The same question may communicate different ideas to different people. For example, "How many people are in your family?" may mean aunts and grandparents to one person and only parents and siblings to another.

Will all students in grades K–8 understand this question?
[They probably will.]

Will all students understand each question in the same way?
[Probably not, since kindergartners generally do not know how to read. The question is likely to mean something different to kindergartners than to fifth graders.]

How can the question be refined to mean the same thing to everyone?
[One possibility is to ask, "Do you like to read or be read to?"]

Will the question get the data you are looking for?
[It looks like it will.]

3. Continue to involve students in activities for writing effective questions. See Illustration 5 for further ideas.

Activity 5: Developing the Individual Survey Question

1. Each student should select a topic and write a first draft of a survey item. Pairs of students should then read and critique each other's questions. Discussing some of the questions with the whole class will

make it clear that every question needs to be carefully evaluated to ensure that it will get the information desired.

2. *Evaluation.* Evaluating students' progress in reviewing and refining their own survey items is important. One way to assess this progress is to have students write their initial question on the top of a page, have their partners write a critique in the middle of the page in a different color ink, and then have the students write the final question below the critique in the original color ink. In figure 1, the teacher first checked whether the partner gave a valid critique of the item. The student did, and therefore received one point. The item-writer was given one point for apparently understanding the critique and another one to two points for appropriately refining the survey item.

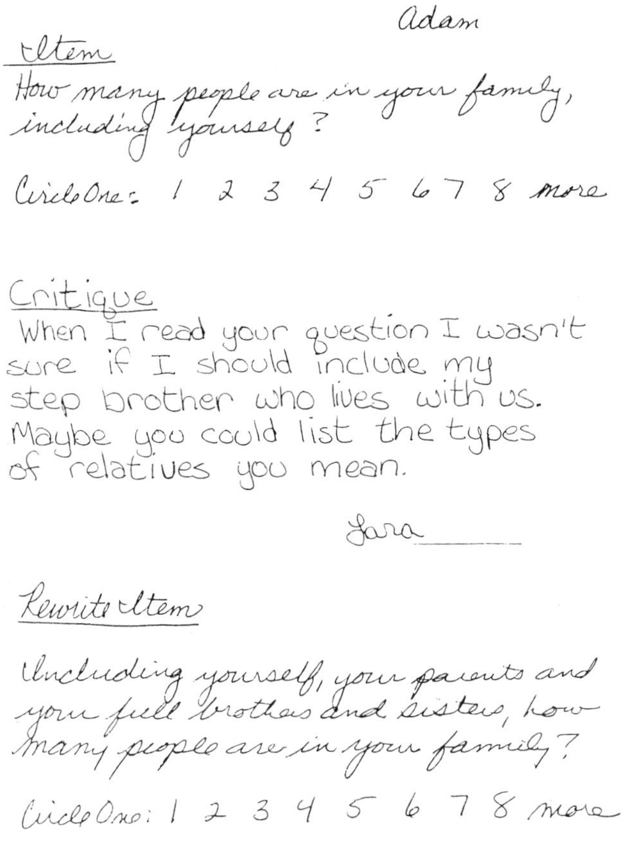

Adam

Item

How many people are in your family, including yourself?

Circle One: 1 2 3 4 5 6 7 8 more

Critique

When I read your question I wasn't sure if I should include my step brother who lives with us. Maybe you could list the types of relatives you mean.

Lara

Rewrite Item

Including yourself, your parents and your full brothers and sisters, how many people are in your family?

Circle One: 1 2 3 4 5 6 7 8 more

Fig. 1

3. Make plans to pilot each question with two or three students in other classes. Discuss the advantages of trying out the questions on a diversity of students (i.e., varied ages or gender). Students should prepare copies of their items, ask potential survey respondents to write their answers, and then ask the respondents what they thought the question meant, why they answered as they did, and whether they thought the question was clear.

COMMUNICATION: Survey items may benefit by first piloting them in interview form. The direct verbal interactions allow students to hear responses directly and to immediately confront issues of miscommunication.

4. After the pilot, students should rewrite their questions in final draft form and have them approved by the teacher. Each student then can enter his or her question on the master survey, which has been set up in a word-processing program on the computer.

USING TECHNOLOGY: One way to randomly select students from a whole school is to assign each student a number, program the computer to generate a finite set of random numbers, and interview students with those numbers.

Activity 6: Whom Should We Interview?

Pose the following questions for discussion:

Whom should we interview?

What would happen if students surveyed only their friends?

What would happen if the survey were conducted in the lunch room when only first- through third-grade students were at lunch?

The students should realize that the above situations would probably result in data that represent only certain subsets of the student body. For example, if they gathered information about height or age during the first- through third-grade lunch, the "average" height or age would be much too low to represent the entire student body. Concepts of bias, sampling, and randomness can be explored in a meaningful context. The best way to avoid sample bias is to use a random sample, which is achieved by striving for a selection process in which all people in the intended population have an equally likely chance of being selected. Often, a random sample is not possible or is too difficult to implement; students need to think seriously about avoiding situations that can result in biased samples and improving the representativeness of a sample. Only after students have encountered these questions in their own collections of data are they able to critique and question information based on other people's sampling and data collection techniques.

Activity 7: Planning the Implementation

Students should plan as a class how the survey will be distributed and collected. Many questions need to be addressed:

Who should administer the surveys?

Has permission been secured from the teachers whose classes would be involved?

When can we take selected students from the teacher's class? Have we developed a schedule?

What if someone does not want to respond to the survey questions?

Brainstorm other possible implementation questions and discuss alternatives for each one. Encourage students to plan for unexpected circumstances. To prepare for the unexpected, students should review the purpose of the survey and be ready to make decisions on the basis of good judgment, sensitivity to respondents' and teachers' needs, and the goal of the survey.

Activity 8: Organizing and Representing the Data

Remind students that they will need to share their outcomes for the class members to develop a "profile" of the "average" student; they will need to put their data in a form that can readily communicate the results to their classmates.

1. Introduce alternative ways to represent data using the information from the five-question class survey developed in Activity 4. Use a variety of plots and graphs on each set of data, discuss the impressions given by each, and decide which ones would be useful in developing a profile of the "average" student. Specific ideas for comparing graphs can be found in Illustration 6: Comparing Plots and Graphs.

2. Writers of the survey items should be given the raw data for their own question. The teacher can cut up the questionnaires and give each

student the part with his or her item. (If a survey-making software program was used, students can readily obtain an individual printout of their data for their own question.) Students should tally and organize the data.

3. Ask students to prepare at least two different plots or graphs appropriate to their data. They should share their representations of the data with the class and discuss the different impressions communicated by the different forms.

Activity 9: Summarizing the Data

It is natural to use concepts of central tendency and distribution when describing and summarizing information to solve a problem or make a decision. Using data gathered by the class furnishes rich opportunities for students to learn about summary statistics in a meaningful context.

1. Use the five survey items developed by the class in Activity 4 to introduce and discuss measures of center and spread appropriate to the data. For example, to develop the concept of the arithmetic mean, divide the students into groups of three or four and ask them to measure their heights. Have each group cut one continuous length of string to represent the sum of their individual heights. Then fold the string into three or four equal parts (depending on the number of students in the group). The resulting length is that of the mean, or arithmetic average. Although this activity uses no numbers, the procedure for finding the mean (i.e., finding the total height and then dividing by the number of people) is clearly evident in the physical actions. Students can compare their own heights to that of the folded string to develop an understanding of why the mean does not necessarily have to be equivalent to any one value in the data set.

REASONING: Using concrete materials rather than numbers to demonstrate procedures can help students develop reasoning skills. Numbers tend to trigger memorized computational responses, whereas the use of concrete materials promotes reasoning procedures.

2. Students should engage in further activities with measures of central tendency and measures of spread. Illustration 7: A Look at the Average Wage compares the roles of the mean, the median, and the mode in a realistic problem setting. Illustration 8: Exploring Standard Deviation is technology-based and requires that students use the statistics functions on a scientific calculator to investigate a real-world problem.

3. Students should summarize their own data and write a brief description of their results to share with their classmates.

Activity 10: Drawing Conclusions and Making Decisions

Finally, the students must use all available data to decide on the general characteristics of the "average" student. Small groups of students can propose prototypes to the class, and students can then discuss how well the prototype represents the student body as a whole. Clearly, there will not be one specific profile that is correct, but there will be many areas of agreement and many discussions about "ranges." After the small-group work and the class discussion, students should be individually assigned the task of writing a poem or a limerick about the "average" student in the school. These poems can be printed in the school newspaper or presented at an all-school show. This task cycles the project back to its interdisciplinary roots.

Closing Comments

Most of these activities may take more than one class period. The project meets the spirit of the *Curriculum and Evaluation Standards* in many ways. The students collect and manipulate their own data, which makes learning relevant and meaningful. The interdisciplinary nature allows students to see how their learning is linked to many topics and processes. Group work encourages students to reason out their ideas as they communicate with and explain to one another. The whole process is challenging and interesting for both students and teachers.

EXPERIMENTS

Student-conducted experiments furnish another opportunity for data collection. This section contains two illustrations; one is a scientific experiment and the other is a probability experiment. In Illustration 2: The Mung Bean Experiment, data are gathered from three different plant-growing conditions. Students use scientific reasoning to draw conclusions from the results. Further, they use many skills typically found in the mathematics curriculum, such as measuring, recording data in tables, and making charts and graphs. The implications of the results also lend this project to interdisciplinary considerations of world hunger and agricultural problems.

Illustration 3: Is This Game Fair? is a probability experiment that appeals to middle school students' enormous interest in fair game playing. At this age, the students' definition of fairness is progressing from an immature concept that a game is fair "if I win" to one in which each player has an equally likely chance of winning. The activities furnish opportunities to collect and organize data, compare and analyze results for two conditions, and draw conclusions about the fairness of the game.

ILLUSTRATION 2: THE MUNG BEAN EXPERIMENT

Tess Jackson and Gary Brooks

CONNECTIONS: Other disciplines provide rich opportunities for students to solve problems by collecting and analyzing data. For example, in agricultural experiments, like this one, students will realize that results can be compromised by roots tangling and breaking during the measuring process. In the social sciences, obtaining accurate counts for population data is very difficult when considering the homeless, students in college, and so on.

This activity strengthens the connection between mathematical analysis and real-world applications through the study of the effects of salt on the growth of mung beans. Mung beans grow rapidly, which allows this experiment to be executed in less than a week. In addition to acquainting the students with the problems inherent in data collection, the teacher can lead the students into the important and relevant issues of world hunger, problems of irrigation with salt water, and the desalinization process.

Materials

Materials per class:

4 one-gallon containers
1 carton of table salt
1 bag of dry mung beans
tap water

Materials per student group:

3 clear 7-oz. plastic cups, each with a plastic grid or screen wire lid
a metric ruler
a cardboard box
scissors and tape

Preparation

Prepare the beans by soaking them for twenty-four hours in a dark gallon container of tap water, changing the water once or twice. After twenty-four hours discard any bean that shows no evidence of germination. Also prepare the three soaking solutions in the gallon containers:

♦ ♦ ♦ ♦ ♦ ♦ ♦ ♦

Treatment 0—no salt per gallon of water (labeled TrT0)
Treatment 1—one teaspoon of salt per gallon of water (labeled TrT1)
Treatment 2—two teaspoons of salt per gallon of water (labeled TrT2)

Activity 1: Setting Up the Experiment

1. Have groups of three or four students prepare three plastic cups with plastic grid or screen wire lids. Each lid should have twenty holes, approximately five mm in diameter and numbered 1 to 20. (See fig. 2.) The students should label each cup TrT0, TrT1, or TrT2, and then fill them with the corresponding treatment solutions. Furnish each group with at least sixty germinated beans and have the students randomly assign twenty beans to each treatment. The initial root length should be recorded for each bean and the bean should then be placed on the grid with the root entering the solution. Use tap water to maintain the solution height and keep beans in a dark box when they are not being measured.

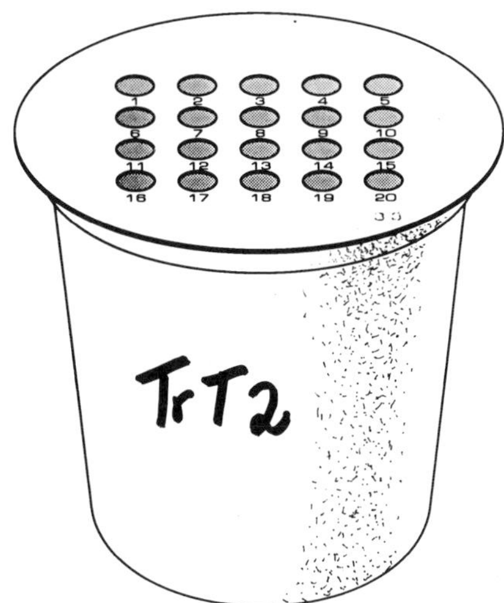

Fig. 2. Array of holes on plastic or screen top

2. *Evaluation.* Students should use a lab manual (bound booklet of graph paper) to record measurements, describe procedures, and note any unusual occurrences. These manuals will help the students as they evaluate their own work, analyze their data, explain unusual results, formulate questions, and draw conclusions. As the teacher periodically reads the manuals, he or she becomes aware of specific skills that need reinforcing or misconceptions that need correcting. Furthermore, at the end of the year, students and teachers together can evaluate their long-term growth in applying the scientific process.

Activity 2: Measuring the Growth

1. At the end of twenty-four, forty-eight, and seventy-two hours, the beans should be removed, one at a time, and the root length recorded. One member of each group should record in a lab manual the root length measurements, the procedures used, all difficulties encountered, and problem resolutions. (Typical growth of TrT0 for each twenty-four-hour period can be found in fig. 3.)

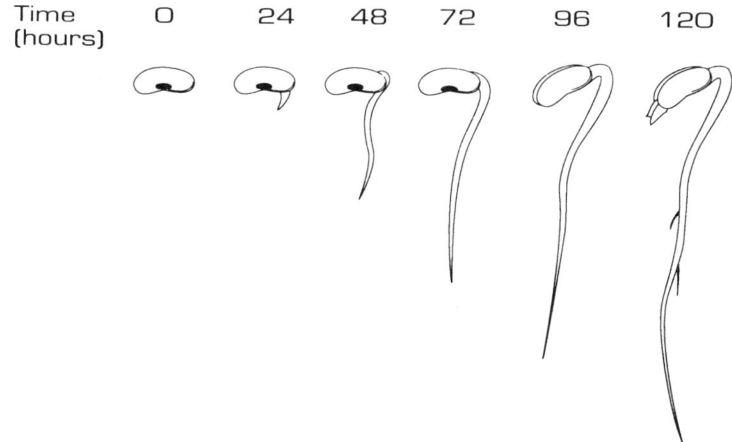

Fig. 3. Growth in each twenty-four-hour period for TrTO

PROBLEM SOLVING: When difficulties occur in data collection, students have to think about ways to deal with the "errors" and the variability. One way is to reconceptualize the design of the experiment and implement it again. Another way is to keep careful records of the problems and account for them in the data analysis by proposing alternative explanations and putting forth hypotheses concerning overestimates and underestimates of effects.

Activity 3: Analyzing the Data

Each group should prepare a descriptive analysis comparing the change of root lengths with respect to time and treatment. The student-gathered data can be used to introduce or review the use of charts, plots, graphs, and summary statistics (including measures of central tendency and dispersion). In their written analyses, students should describe the procedures used and discuss the problems encountered and their resolutions. For example, students will probably encounter roots that curl and are difficult to measure, roots that break off and can no longer grow, and beans that fall through the holes and cannot be identified. The written notes taken during the laboratory periods will be very valuable as the students interpret and analyze their data.

Closing Comments

Much of the field of probability and statistics developed as a result of scientific experimentation (especially in agriculture), so it is very natural to blend the scientific process into the study of statistics. This illustration exemplifies not only connections (to other disciplines), but also problem solving, reasoning, and communication.

ILLUSTRATION 3: IS THIS GAME FAIR?

Hope Martin

This illustration begins with a problem situation in which students are asked, "Is this game fair?" The game is designed for two students, each of whom starts with ten points. One student is the "player" and the other is the "opponent." Only the player rolls the number cubes. Each time the player rolls a sum of 7, the opponent must give up, or transfer, three points to the player. When the player rolls a sum other than 7, he or she must give up, or transfer, one point to the opponent. The winner has the most points at the end of ten rolls. A participant who runs out of points before ten rolls loses immediately. When they are familiar with the rules, students are asked to predict whether or not the game is fair. They will decide whether or not their conjecture is correct in two ways: by gathering enough data to be convinced and by using their understand-

◆ ◆ ◆ ◆ ◆ ◆ ◆ ◆

ing of probability and odds to analyze the game mathematically.

Materials

2 six-sided number cubes for each pair of students
calculators
blackline masters 1 and 2

Preparation

Photocopy a classroom set of blackline master (BLM) 1.
Make overhead transparencies of BLMs 1 and 2 and figure 4.

Activity 1: Understanding the Game and Making a Prediction

Give students a copy of BLM 1. Familiarize students with the game by reviewing the rules at the top of BLM 1 and playing one game, with the teacher as the player and the class as the opponent. Record the results on the overhead transparency to illustrate how students will need to record their own results on BLM 1. Then ask students to predict *in writing* whether or not they think the game is fair. They can write their prediction and reasons on the back of BLM 1.

REASONING: Making predictions, gathering information, and modifying the prediction on the basis of new information are fundamental to developing higher-level reasoning. Students need more experiences in which the correctness of their answers is not immediately evident; in which they need to gather further information; in which they are encouraged to reconsider their original prediction; and in which they feel comfortable modifying and refining their initial conjectures.

Activity 2: Playing the Game and Gathering Data

1. Put students into teams of two, an opponent and a player. Give each pair of students two number cubes and one copy of BLM 1 to keep track of the points in the game. Student pairs should decide who will be the player and who will be the opponent. Only the player rolls the number cubes.

2. Using BLM 1, review the rules of the game again and help students record the results of their tosses. Each pair of students should play ten rounds, recording their results after each toss.

Activity 3: Who Won?

1. Although some teams will have opponents as winners and some will have players as winners, the students can note trends by collecting all the class data on an overhead transparency of BLM 2. As each team reports its winner, fill in the cells on the transparency. Use calculators to determine the percentage of games won by the opponents and won by the players.

2. Pose these questions for discussion:

Do you feel more strongly about your prediction? Why?

Do you feel less convinced about your prediction? Why?

Do you want to modify your prediction? Why?

Have students summarize their revised convictions on the back of BLM 1.

3. If the data are not yet convincing (that the opponent has an advantage in the game), have students play another round and then combine the new information with the old. Repeat until all students are fairly well convinced that the opponent has an advantage. Students should be given the opportunity to reflect on the above questions and modify their predictions after each round.

Activity 4: Why Does the Opponent Seem to Win?

1. Once students have become convinced from the data that the oppo-

Sums of Seven

+	1	2	3	4	5	6
1						
2						
3						
4						
5						
6						

How many different sums can we get when we roll a pair of number cubes? Let's use this grid to determine all possibilities.

Fig. 4

PROBLEM SOLVING: A valuable way to help children "look back" and notice "similar problems" is to modify a problem just solved and ask students to evaluate the new situation. Explicit discussion on the similarities and differences between the original and new situations helps students learn what it means to thoroughly analyze a problem.

nent has a distinct advantage in this game, it is time to analyze the game mathematically. Use a transparency of figure 4 to investigate the various ways that sums can be made on two number cubes.

2. Once the class has helped you complete the table, point out that the probability of rolling a sum of 7 is 6 out of 36, or 1/6. Now, consider the *odds* in favor of rolling a sum of 7 (1 to 5, or 1:5). Given this information, ask students how the game could be changed to make it a fair game. (Since the odds in favor of rolling a sum of 7 are 1:5, the opponent should give up five points to the player for each sum of 7, instead of three points.)

Activity 5: Extensions

1. Teams can play the game again with the new five-point rule. Gather data to verify that this rule does indeed make the game fair.

2. Play the same game, but when the player rolls a *double*, the opponent gives up three points to the player; when the player rolls anything other than a double, the player gives up one point to the opponent. Ask for predictions and gather data again. (This game is essentially the same as the original game, since the odds in favor of rolling a double are 1:5.)

Closing Comments

Probability is a *measure* of chance. Before students deal with the formal mathematics of probability, they need to understand the meaning of chance and have some experiences in naturally quantifying the concept of likelihood. This illustration lets students explore the concept of equally likely chances and quantitatively describe chance in various situations. Students frequently say, "The opponent is more likely to win" or "The player's chances of winning are not as good as the opponent's." These intuitive ways of applying quantity to the level of chance are building the intuitions needed for understanding the concept of probability. More probability experiments can be found in Phillips et al. (1986).

SIMULATIONS

Many interesting problem situations are too difficult for middle school students to analyze mathematically, or they may be inconvenient, impossible, or too dangerous for experimentation. In such circumstances, *mathematical models* are used to simulate the problem setting and generate data. The Monte Carlo method of simulation uses random number devices such as dice, number cubes, coins, spinners, or lists of random numbers (from tables or computer random-number generators) to represent the mathematical characteristics of the real-world situation. An experiment is designed to *model* the problem situation, and simulated data are generated and analyzed as though they were real data. The following illustration provides an in-depth example of this third type of data collection activity for middle school students.

ILLUSTRATION 4: MONTE CARLO SIMULATIONS

Marsha Landau

The theoretical basis for the Monte Carlo method of simulation is the law *of large numbers:* as a simulation is run more and more times, the simulated estimate

$$\frac{\text{number of successes}}{\text{number of runs}}$$

becomes approximately equal to the theoretical probability (Watkins 1981).

Monte Carlo simulations make many interesting, real-world problems accessible to middle school students with no previous instruction in probability and statistics. Two basic types are (1) problems that involve determining the probability of a success or of a failure and (2) problems that ask for an expected value. For probability problems, the simulated data are used to estimate the theoretical probability in the following ratio:

$$\frac{\text{number of successful runs}}{\text{number of runs}}$$

For expected-value problems, the simulated data are used to estimate the expected value by using the following ratio:

$$\frac{\text{sum of results from all individual runs}}{\text{number of runs}}$$

The example below is an "expected value" problem that middle school students find interesting but cannot solve analytically. Further, the problem setting is too expensive and inconvenient to experiment with directly in the real world. A Monte Carlo simulation is developed and used to analyze the situation.

Materials

six slips of paper, a hat, dice, six-section spinners
other random number devices (optional)
a computer programmed to generate random numbers (optional)

Preparation

Write this problem on the chalkboard or an overhead transparency:

> The Tripl-Bubl Gum Company decides to promote its gum by including in each pack the photo of one of six rock music stars. Assuming that there are equal numbers of photos of each of the six stars and that when you buy a pack of gum your chances of getting any of the six photos are the same, about how many packs of gum would you expect to have to buy to get all six photos?
> (Adapted from Travers and Gray 1981, p. 327)

Activity 1: Present the Problem and Make Conjectures

1. After they have read and understand the problem described above, ask students to make some guesses. Some students typically guess 12, 36, and other numbers in response to the problem but have no justification for these answers. Other students point out that the answer could be 6, for a very lucky purchaser, and that it is possible to buy 100, 1000, or even larger numbers of packs of gum without completing the set. At this point the students are ready to take another look at the problem.

2. Pose these questions for discussion:

About how many packs would you expect to have to buy?

It is possible to get all six photos with only six packs, but would you expect that to happen?

It is possible not to get all six photos with 100 packs, but would you expect that to happen?

COMMUNICATION AND REASONING: Even when students make reasonable estimates for situations, they are often unable to justify their conjectures. Encouraging students to verbalize their thinking helps build their reasoning skills. The process of verbalizing reasons and justifications forces students to better evaluate and organize their own thinking. As their ability to communicate with others improves, so do their internal thinking processes, helping them develop higher-level reasoning skills.

Do you think the problem is realistic?

Has anyone had an experience collecting objects in offers like the one described?

If you knew that you would need to buy about 100 packs of gum to obtain all six cards, would you still attempt to do it? [Students should be helped to realize that there may be more economical ways to obtain the whole set of photos.]

Activity 2: Modeling the Problem

1. The situation appears in the form of a problem-solving activity. One problem-solving strategy appropriate for this situation is to "act it out." However, the class is not about to go on a field trip to the supermarket. The challenge is to find a way to "act it out" in class using available materials. To consider ways to simulate the problem in the classroom, students must first consider the essential mathematical characteristics of the situation:

Each pack contains one of six rock music star photos.

There is one chance in six that any particular photo is in a particular pack of gum.

Duplicates are possible.

2. Ask students what could be used instead of packs of gum to help *act out*, or *simulate*, the problem situation. One suggestion can be to put six slips of paper, each containing the name of one rock star, in a hat. Buying a pack of gum is simulated by drawing a slip; the name is recorded, the slip is replaced, and the drawing is repeated until all six names have been recorded. Then count the number of names that have been drawn, for one *trial*.

3. Run one trial of this simulation with the students. Suppose that twenty drawings were necessary. Pose these questions:

Would you now expect *to buy twenty packs of gum to obtain all six photos?*

If you run another trial, do you think you will need twenty draws again?

How can we get a result we "believe in"? [Run more trials.]

Are there other materials we can use to act out the purchases?

The "slips in a hat" model guaranteed that, on each draw, the six names were equally likely outcomes. The probability of drawing each name was 1/6. Any device that incorporates six equally likely outcomes to simulate each purchase of a pack of gum could be used. For example, we could use a six-sided die, a spinner with six equal sectors of a circle, six distinct playing cards in a brown paper bag, and so on. We could also use random numbers from a table of random numbers or those generated by a computer.

Activity 3: Running the Trials

The simulation is most fun when several random number devices are available for each pair or group of students. Once the devices have been selected, the next step is to find an efficient way to do the record keeping. Elicit suggestions from the class. Many recording schemes are possible, but the best method is probably listing the six names (or numbers) and putting a tally mark after each name when it is drawn. See figure 5 for an example.

	Trial #1	Trial #2	Trial #3
Rock Star #1	ǀǀ	ǀǀ	𝍸 ǀ
Rock Star #2	𝍸	ǀǀ	ǀǀ
Rock Star #3	ǀ	𝍸	𝍸
Rock Star #4	ǀǀǀ	ǀ	ǀǀǀǀ
Rock Star #5	ǀǀ	ǀ	ǀ
Rock Star #6	ǀǀǀǀ	ǀǀ	ǀǀǀ
Total number of picks	17	13	21

Fig. 5

Be sure to stop when all six names or numbers have been drawn, then add the tally marks to find the total number of picks. We are interested in the total number for each trial. How many trials should be run? As many as possible! In a very short time the students can run at least twice as many trials as there are students.

Activity 4: Combining the Class Data

With many groups generating data, the students might naturally ask how they can use all the results to obtain one believable answer to the problem. If no one suggests finding the mean, ask leading questions to get the students to recognize that they need *one* number to represent many numbers. Organize the class data in the following manner:

1. Write each of the sixty or so data points on the board.
2. Next, create a frequency distribution.
3. From the data displays above, identify the range of values, the median, and the mode.
4. Use a calculator to find the mean and compare the mean to the two other measures of central tendency—the median and the mode. [It should be around 15.]
5. Decide as a class how many picks should be expected.
6. Compare the results to the original conjectures.
7. Decide whether it would be "worth it" to buy fifteen packs of gum in order to collect a set of six photos. (See Lappan and Winter [1980, p. 448] for the calculation of the theoretical expected value of 14.7.)

Activity 5: Looking Back

1. Discuss ways to improve the estimate. [Run more trials.]

2. Were the results different when using different models? Why? (For example, many children hold down a corner of the spinner and thereby alter the "fairness" of the device; some dice may not be perfectly balanced; and so on.)

3. Extend the problem by considering how the solution would change if there were eight rock music star photos instead of six. What models

◆ ◆ ◆ ◆ ◆ ◆ ◆ ◆

could be used? (Show the students that there are four-, eight-, ten-, twelve-, and twenty-sided dice; some students may have seen these dice in adventure or fantasy games.)

USING TECHNOLOGY:

Computer simulations are very powerful and useful simulation tools. Middle school students can use computer simulations successfully once they have had experiences conducting simulations with physical devices. See Collis (1982, p. 585) for a computer simulation program in BASIC.

Activity 6: An Extension Problem

1. Pose the following problem to students:

> The Kid has challenged Doc to a pie-throwing contest. Both have bad aims. The probability that Doc hits the Kid on any one throw is 1/10; the probability that the Kid hits Doc on any one throw is 1/5. Doc really would not like to get whipped cream on the fancy tie he is wearing, so he thinks that throwing pies might be a bad idea. "Okay," says the Kid, "to even things out, you can go first and we'll alternate throws until one of us gets hit." Doc wonders if this really evens things out. (Adapted from Watkins 1981, p. 204)

2. Small groups of students should devise a plan to simulate this problem. (Hint: One method is to act out the pie-throwing contest using a table of random numbers or the list of the last four digits of phone numbers in a telephone directory. Let the Kid pick a number 0 through 4. If the Kid picks 3, then on the Kid's turn, 3 is a hit; 0, 1, 2, or 4 is a miss; and all other digits are disregarded. Doc picks a number 0 through 9 for himself. Starting at a random location in the table, Doc lets each number represent a pie toss. If it is his turn to throw, 5 will represent a hit and any other digit, a miss. Repeat the procedure many times.)

3. Each group should run enough trials to draw and defend a conclusion. It will be clear after a number of trials that the Kid has a distinct advantage in this problem.

4. *Evaluation.* After students have completed a few simulation problems, ask them to reflect individually on and answer in writing questions such as the following:

PROBLEM SOLVING.

Problem-solving strategies such as "consider a similar problem" and "look back" are most effectively taught when you help students recognize when they have naturally used such procedures. Providing opportunities for students to reflect on their own processes and share their examples is an effective means of illustrating these problem-solving strategies to students.

Did you (or your group) use ideas from other simulation problems when developing a plan for simulating this problem? Explain how.

Did you (or your group) come up with alternative methods for simulating the problem? Explain what they were and why you decided to discard or use them.

After you came up with your answer, did you (or your group) consider whether it made sense? Describe how you thought about this.

Evaluation of students' responses can provide a basis for class discussion on students' own problem-solving processes. Ask students to share their specific experiences in relating the new problems to ones they had previously completed, in considering alternative methods, and in looking back at their solution.

Closing Comments

Let the students know that they have been doing Monte Carlo simulations in which random numbers were generated by devices, such as dice and spinners, that are sometimes used in gambling. Monte Carlo methods are especially useful for solving problems in which calculating the result theoretically is extremely difficult or beyond the mathematical background of the students.

Finally, give students further opportunities to design and carry out simulations. Small-group work with further simulation experiences will help children grow in their ability to use mathematics to model real-world situations. Additional examples can be found in Choate (1979, p. 41), Simon and Holmes (1969, p. 283), Travers (1981, pp. 215-16), and Watkins (1981, pp. 207-8).

CHAPTER 2
A FOCUS ON COMMUNICATION

Communication issues are central to the study of statistics. The process of collecting data often involves communicating, since many surveys and opinion polls are based on questions posed to people. Further, data are usually collected to form arguments, make decisions, or develop summary descriptions. All these purposes involve communicating conclusions implied by the data. Although this section focuses on communication issues, the illustrations that follow characterize many other aspects of mathematics promoted by the *Curriculum and Evaluation Standards*, such as problem solving, reasoning, and making connections.

WRITING SURVEY QUESTIONS

The following illustration focuses on a small portion of a long-term survey project, similar to Illustration 1. As part of long-term survey projects, students need to develop techniques for writing effective survey questions to use in their individual projects. The following illustration provides activities that help students become aware of important issues and develop their own abilities to analyze and write survey questions.

ILLUSTRATION 5: WRITING SURVEY QUESTIONS

Lynn Dinkelkamp

Preparing a survey truly integrates language experience with mathematics. Not only must students write questions that pertain to a particular topic, but also they must be able to effectively use these questions to obtain quantitative data that can be analyzed. Finally, students must be able to communicate some conclusions. As students write and implement their own survey questions, they begin to appreciate how important the question statement is in obtaining the quality and the type of information they want. The following activities are designed to be used as students prepare questions for their own survey projects.

Materials

blackline masters 3 and 4

Preparation

Reproduce classroom sets of blackline masters 3 and 4.

Activity 1: A Professionally Prepared Survey

Furnish each student with the survey on BLM 3, which was adapted from a survey prepared by the Bureau of the Census, Department of Commerce, Washington, D.C. Ask students to complete the survey, and then pose the following questions for discussion.

Were all the questions clear to you? If not, tell why.

Did you always know what to answer?

Do you think everyone else answering the questions would know what to answer? Why or why not?

What type of responses were required in the survey? Multiple choice? Numerical value? Short answer? Fill-in-the-blank?

Why do you think they used the types of responses they did?

What do you think the data will look like when it is combined and organized?

What questions do you think the writers of the survey are going to try to answer with the information they obtain?

How do you think the data will be used?

Activity 2: Survey Questions without Bias

Tell students that a Chicago radio station took a poll to find out whether the people in the metropolitan area preferred the White Sox or the Cubs baseball team. The people polled were asked these two questions:

a. On a scale of 1 (low) to 10 (high), how well do you like the great White Sox baseball team now that they are winning the division?

b. On a scale of 1 (low) to 10 (high), how well do you like the Cubs baseball team?

Small groups of students should be assigned to answer the following questions about the survey items, which will then be discussed with the whole class:

Will the resulting data be easily tallied and organized?

What kind of results would you expect?

Will the results accurately reflect the comparative feeling people have for the White Sox and the Cubs?

Can you make a conjecture about which team the question writer prefers? Why do you think so?

Can you think of any guidelines that should be followed in writing effective survey questions? What are they?

Students quickly recognize that using certain words (like *good* or *poor*) and giving extra information can influence people to answer the questions differently from what they really think. The following guideline was developed by students in a seventh-grade class: Try to avoid including words or extra information that may influence or change the way a person thinks. Surveys are usually conducted to find out how people think on their own.

Activity 3: Writing Well-defined and Clear Survey Items

Prepare a worksheet that has pairs of potential questions for a survey. (Examples are given below.) Students are to pick the item that they would prefer to use on their own survey, defend their choice, and try to develop guidelines for writing effective survey questions.

Pair A: 1. How many brothers and sisters do you have? __________

2. Are you a member of a large family? yes no

As noted in Illustration 1, students readily realize that "large family" can be interpreted in different ways. One seventh-grade class developed the following guideline for writing survey questions: Make sure you use terms that everyone will understand in the same way.

Pair B: 1. Should the explosion in today's technological capabilities have an impact on school curriculum involving applications in mathematics? yes no

***PROBLEM SOLVING:* As an activity in problem formulation, give students some written conclusions that were based on a survey. Ask students to make conjectures about what questions may have been asked to produce such conclusions. Then compare their questions to the ones actually used.**

***CONNECTIONS:* Interdisciplinary connections between mathematics and other academic areas need to be made throughout the middle school experience. Otherwise, students begin to believe that quantitative concepts and techniques taught in mathematics class are only useful and valid in mathematics class. Student-conducted surveys are one means for making interdisciplinary connections.**

2. Should school-age children be allowed to use calculators for problem solving in mathematics class? yes no

Most students will notice that the first question is very long and not easy to understand, whereas the second is shorter and more straightforward. A group of middle school students developed this guideline: A survey question should be easily understood and not too long.

Pair C: 1. Should elementary school or high school students be allowed to use calculators on all their mathematics assignments?

2. Should calculators be allowed on all mathematics assignments in elementary school? yes no
In high school? yes no

Students will quickly notice that a person may feel differently about elementary school and high school uses of calculators, so a better survey item is the second one. Students may create this guideline: A survey item should ask only one question. The person responding should never feel that there may be two different answers.

Activity 4: Improving First Draft Survey Questions

Give a copy of BLM 4 to each student. Each survey item is in first draft form and has room for improvement. As homework, the students should write an improved version of each item and tell why they made the change. The changes should be discussed in small groups the next day, and the group should turn in one set of agreed-on changes and explanations for the changes.

Activity 5: Running a Mock Survey

Divide the students into small groups. Choose a topic of investigation and briefly outline six questions to ask about the topic. Half the groups should purposely cast the questions in a biased form and the other half should cast the questions in a nonbiased form. As a class, decide on six biased questions for the mock (or biased) survey, and six parallel questions for the nonbiased survey. Every student should interview two people, one with the mock survey and the other with the nonbiased survey. The following day, the results should be combined and compared in terms of parallel questions.

TECHNOLOGY: Using a computer, a survey-making software program, and an overhead projection of the monitor can facilitate the data collection and representation process, allowing for fairly quick comparisons between the two sets of data. Thus, more time can be spent on dealing with the issues of recognizing and writing more effective survey items.

Closing Comments

Middle school students can become critical consumers of survey information when given opportunities to experience all phases and aspects of the complete survey process. These activities for writing survey questions are very effective with middle school students, because they begin to realize the important issues of communication in data gathering.

REPRESENTING DATA

Communication is achieved in many forms—oral, written, and pictorial, and so on. When plots and graphs are used to represent data, the pictorial displays allow for visual comparison of data and for analysis of mathematical relationships that often cannot be easily recognized in written numerical form (Curcio 1989). Further, as noted in the *Curriculum and Evaluation Standards* (NCTM 1989), different visual presentations convey different perspectives. The choice of the form depends on the questions to be answered and the impressions to be communicated.

♦ ♦ ♦ ♦ ♦ ♦ ♦ ♦

Developing Graph Comprehension (Curcio 1989) includes an activity in which students estimate and then count the number of raisins in a 1/2-oz. box. The results are presented in a variety of visual forms, and then the information conveyed by each representation is compared and contrasted. The following illustration, adapted from Curcio (1989), assumes that students have previously been introduced to back-to-back stem-and-leaf plots, double bar graphs, and box plots.

ILLUSTRATION 6: COMPARING PLOTS AND GRAPHS

In this activity, students are asked to estimate the number of raisins in a 1/2-oz. box. After guessing the number of raisins, the students count the actual number in each box. The data from the estimates and the actual counts are represented in several visual forms. The focus of this illustration is to compare and contrast the information that is communicated by three representational techniques: the back-to-back stem-and-leaf plot, the double bar graph, and the box plot.

Materials

one 1/2-oz. box of raisins for each student
blackline master 5

Preparation

Reproduce a classroom set of BLM 5.

Activity 1: Gathering and Representing the Data

1. Give each student a 1/2-oz. box of raisins. Tell them not to open the boxes. Ask, "How many students have eaten raisins from a box like this?" After they raise their hands, ask, "How many raisins do you think you get in one of these boxes?"

2. Record student responses on the chalkboard or overhead projector in a stem-and-leaf plot, like the one shown on BLM 5. (Note: So far, you will have "leaves" only on the right-hand side.)

3. Now ask students to count and report the actual number of raisins in each box. Record their results on the left-hand side of the stem-and-leaf plot.

4. Ask students to work in groups of three or four to develop at least two more visual representations of the class data. Each group should present its graph or plot to the class.

Activity 2: Interpreting Three Visual Representations

1. Pass out copies of BLM 5 to students. Tell students that these data were gathered by another class when comparing estimates to actual counts of raisins. Ask students to examine the back-to-back stem-and-leaf plot on BLM 5, and then pose questions such as the following:

What is this plot about?

Which actual count occurred most frequently (i.e., the mode)? How can you tell by looking at the graph?

How can you find the lowest actual count on this plot?

Which estimate is the lowest? The highest?

How can you find the range of the estimates on this plot?

USING TECHNOLOGY: New advances in technology have resulted in software packages that construct plots and graphs for given data sets. Curcio (1989) recommends that once students have developed an understanding of different graph types, they can use the computer as a powerful tool to focus on analyzing similarities and differences in the different forms.

What is the range of the actual counts?

How can you use the plot to find the median of the estimates?

What is the median of the actual count?

What do you notice about the estimates compared to the actual counts? (For example, the estimates were more spread out than the actual counts.)

What is being compared on this display?

How does this data compare to our class data?

Was our class better at estimating than they were? How can you tell?

Evaluation. As you progress from factual questions to questions that require higher order thinking, you might find it helpful to have students write their responses. Reading students' responses to questions such as the last one above helps teachers evaluate whether or not students are learning to use representational techniques as tools in decision making. For example, a student's response indicates a high level of understanding:

> Students in our class tended to guess too high, while students in the other class seemed to guess too low. I could tell because the bulge on our estimate side was next to numbers that were higher than the bulge on the actual count side. Students in the other class didn't have as big a bulge with their estimates, but their estimates kind of stuck out at lower numbers than on the actual side.

2. Ask students to examine the double bar graph. Point out that this is only part of the graph, since it displays the specific estimates and counts of only five students. Pose questions such as these:

What is this graph about?

Which estimate is the lowest among these five students? The highest?

Which actual count is the lowest among these five students? The highest?

Which actual count occurred most frequently among these five students (i.e., the mode)? How can you tell by looking at the graph? [Estimate from the height of the bars.]

How can you tell which estimate occurred most frequently?

What is the range of the estimates? The actual counts?

What is the median of the estimates? How can you tell from the graph?

How can you find the median of the actual counts from this graph?

What is being compared in this graph?

What would the graph look like if the data from all twenty-five students were included? [It would have twenty-five double bars. But we can't actually draw them with the available data.]

If we had full double bar graphs from both classes side by side, what would be similar? What would be different? How can you tell?

3. Ask students to examine the multiple box plot on BLM 5. Pose questions such as the following for discussion:

What is the plot about?

From the box plots, can you tell which estimates occurred most frequently? Why?

From the box plots, can you tell which actual count is the lowest? Explain.

Can you tell what the lowest estimate was? The highest?

How can you estimate the medians from the plots?

What was the median from the actual count?

What is being compared by these two box plots?

Compare our box plots to theirs. Can you tell which class tended to have the better estimates? How can you tell?

Activity 3: Comparing the Representations

On the chalkboard or overhead projector, put up a heading for each of the three visual displays. Under each heading, elicit from the students the types of information that are and are not readily communicated by each type of visual representation. Write these features in the chart as seen in figure 6.

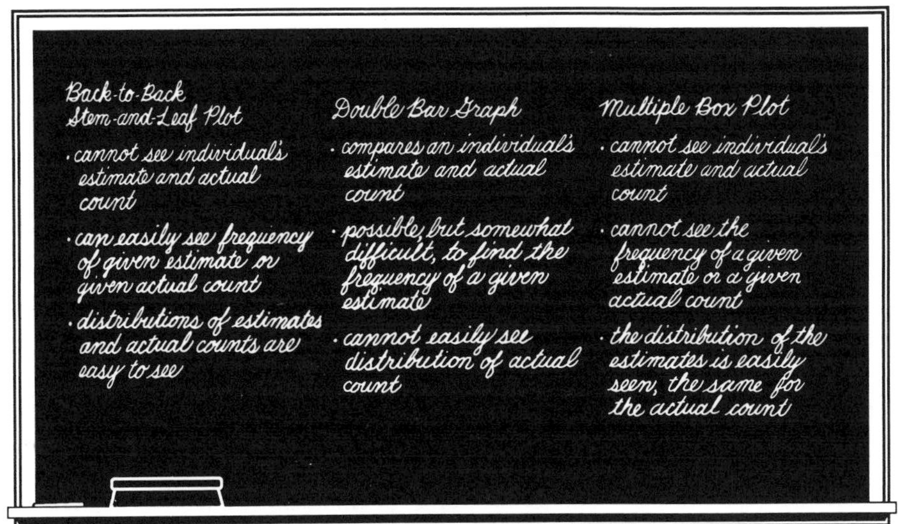

Fig. 6

The following questions can help students think about the different impressions conveyed by the various visual presentations:

What is the difference between the bar graph and the stem-and-leaf plot? What is the same?

What is the same about the stem-and-leaf plot and the box plot? What is different?

What is the difference between the bar graph and the box plot?

What information can you obtain from a box plot that you cannot get from a bar graph? Compare other graphs in this way.

What questions might be answered by each type of presentation?

How should you decide which representation to use for a set of data?

◆ ◆ ◆ ◆ ◆ ◆ ◆ ◆

Activity 4: Comparing Representations of Your Own Data

Ask students to consider the features listed in the chart from Activity 3. Ask whether the same features still hold when they consider the data representations from their own class. If students in the class used other representations, such as pie charts or histograms, expand the chart of features to include these additional types of graphs and plots.

Evaluation. Quizzes and homework assignments should include questions that require students to use higher-order thinking and reasoning. A sample quiz with three questions follows:

Name________________________________

Circle the visual representation(s) that can be used to answer each question. Briefly explain your choice(s). You can use blackline master 5 from class as a reference.

1. Did students in the class tend to overestimate or underestimate the actual number of raisins in their boxes?

   STEM-AND-LEAF PLOT   DOUBLE BAR GRAPH   BOX PLOT

   Explain:

2. What estimate was made most often?

   STEM-AND-LEAF PLOT   DOUBLE BAR GRAPH   BOX PLOT

   Explain:

3. What was the median number of raisins actually in the box?

   STEM-AND-LEAF PLOT   DOUBLE BAR GRAPH   BOX PLOT

   Explain:

Closing Comments

Many middle school students believe that learning statistics consists of mastering specified techniques and skills for constructing stem-and-leaf plots and bar graphs. However, the *Curriculum and Evaluation Standards* makes clear that even more important are the abilities to understand and interpret the various representations, to understand the similarities and differences, and to make informed decisions about which one(s) to use in communicating a complete and honest picture for sound decision making.

CHAPTER 3
A FOCUS ON PROBLEM SOLVING

So far, every illustration in this book has involved problem solving. A fascinating feature of learning about data and chance in the spirit of the *Curriculum and Evaluation Standards* is that an investigative approach is inherently a problem-solving approach. The mathematical development of statistics and probability has been largely motivated by their real-world applications. The focus on this section is on problem settings from which standard and nonstandard curriculum topics can be taught.

PROBLEM SETTINGS FOR STANDARD CONTENT

Measures of central tendency (mean, median, and mode) are commonly taught by defining and then applying them to small sets of numbers. A more effective method of introducing the three measures of central tendency is in a decision-making, problem-solving environment. Illustration 7: A Look at the Average Wage assumes that students have learned how to find the arithmetic average, or mean, but the median and the mode can be first introduced in this context.

ILLUSTRATION 7: A LOOK AT THE AVERAGE WAGE

Albert P. Shulte

The scenario for this illustration is a manufacturing and marketing company, in which the notion of "average" wage is considered from different points of view. The purpose is to show how the selection of the mean, the median, or the mode gives different answers to the same question.

COMMUNICATION: Sometimes everyday use of words may be subtly different from the way the same words are used in mathematics. For example, many people do not realize that an "average" may be represented by the median.

Materials

blackline masters 6 and 7
calculators for students
spreadsheet software and a computer (optional)

Preparation

Reproduce a classroom set of blackline masters 6 and 7.

Activity 1: Presentation of the Scenario

Distribute copies of BLMs 6 and 7. Review with students the problem setting and the salary information. Ask students to identify how the mean, the median, and the mode are used in the problem description. Use question 1 on BLM 7 to review one way in which pay raises may be distributed.

Answers	a. New median:	\$15 000
	b. New mean:	\$23 625
	c. New mode:	\$15 000

Pose these questions:

Which measures of central tendency stayed the same?

Which measures of central tendency changed? Why?

If you changed only one or two salaries, which measure of central tendency will be sure to change? [The mean, since its calculation includes all values]

If you changed only one or two salaries, which measure of central tendency would be most likely to stay the same? [The mode is most likely to stay the same because it is the most frequently occurring salary, and only one or two salaries are being changed.]

If you change only one or two salaries, how likely is the median to change? [It depends. If the median is embedded in the middle of several salaries that are the same, it won't change. If the median is close to a different level of salary, it is likely to change.]

TECHNOLOGY: Using spreadsheet software with this illustration problem is very helpful. Manipulating every number, even with a calculator, can be tedious. The spreadsheet recalculates all the data at once and provides immediate results. Thus, you can focus on the effect of making certain alterations that may otherwise be lost in the quagmire of individual computations.

Activity 2: Using a Spreadsheet (optional)

Have students enter the employee salaries into a spreadsheet on the computer. Column A could list the number of employees of each type, and column B could list the salary of that type of employee. Display the mean salary for all employees in a cell at the bottom of the spreadsheet, labeled "mean salary." (Define the cell as the total salary value divided by the total number of employees.) Using the spreadsheet to display the new salaries and calculate the new mean for each situation, pose the following inquiries:

Predict the mean if the twenty-four lowest paid employees have their salaries increased to $15 000. Make the change in the spreadsheet to find the actual mean.

The president gave himself a raise that resulted in increasing the mean salary $1000. Predict what you think his new salary was. Use the spreadsheet to experiment and find the new salary.

Two new employees were hired by the company: a plant manager and a foreman. Predict whether the mean salary will increase, decrease, or stay the same. Explain your prediction. Check it out with the spreadsheet.

Activity 3: Developing an Argument

Use question 2 on BLM 7 to initiate a discussion on drawing conclusions from the information. Small groups of students can develop position statements and report back to class. There is no single correct answer to the discussion question. Management would naturally favor the mean; the union leader, the median; and the lower-paid union members, the mode.

Evaluation. This problem has more than one reasonable solution. However, many middle school students expect problems in mathematics class to have only one correct solution. Teachers can promote student consideration of multiple solutions by asking students to write up or present at least two reasonable alternatives. At first, students may simply take ideas from one another without much reflection, but as the school year progresses and the teacher continues to value creative, reasonable alternatives, students will begin to enjoy actively looking for multiple solutions.

Activity 4: Evaluating the Measures of Central Tendency

Students and teacher should bring in newspaper articles and other sources of data summaries from the real world. Often the median is reported instead of the mean, especially when extreme values are in the data sets. For example, housing costs are usually reported as medians because a small number of homes in the United States are extremely expensive. In this case, the median better represents the central tendency of home costs than does the mean. By using a variety of contexts and examples, teachers can help students develop general guidelines about when to use the mean and median to represent data sets. Further, by noticing the mode in a variety of contexts, students will find that it is not as stable as the median, and thus the median is usually the better predictor of the center of the distribution.

Closing Comments

The *Curriculum and Evaluation Standards* encourages teachers to go beyond simply defining terms and applying definitions to small sets of numbers. For students to gain powerful decision-making tools in mathematics, they must learn and use the mathematical ideas in contexts that provide rich understandings. This illustration exemplifies how one topic in statistics can be taught through a problem-solving setting.

CONNECTIONS: When students learn mathematical ideas (such as mean, median, and mode) in isolation from context, they soon forget the meanings. When students learn mathematics as it is connected to real-world settings, they are more able to recognize and apply the concepts to new situations.

PROBLEM SETTING FOR NONSTANDARD CONTENT

The increasing availability of scientific calculators in the middle school mathematics classroom furnishes new opportunities for students to learn and use sophisticated concepts. One such topic is standard deviation. Students in Japan and Europe study standard deviation much earlier than students in the United States. Now, using the scientific calculator, we can introduce this topic in the middle school. In the following illustration, the concept of standard deviation is explored by using the statistics mode on the scientific calculator. Using the calculator frees students from the burden of lengthy computation as they quickly find the standard deviations, compare and contrast the results, and use inductive reasoning to draw conclusions and implications for real-life situations.

ILLUSTRATION 8: EXPLORING STANDARD DEVIATION

Barbara Wilmot

The data for this illustration were collected and organized by a middle school class. Middle school students can replicate the long-term experiment to collect their own data for analysis. The focus of this illustration is on description, interpretation, inference, case making, and decision making.

Materials

blackline masters 8 and 9
scientific calculator for each student

Preparation

Show students how to use the scientific calculator's statistics mode. Prepare classroom sets of blackline masters 8 and 9.

Activity 1: Understanding and Describing the Problem Setting

1. Distribute BLMs 8 and 9 to the students. Introduce the problem setting: The light bulbs from three different companies were being compared by randomly selecting bulbs from the three brands and testing them for longevity.

Ask students what question(s) the data collection and analysis were designed to answer.

Ask students how they can use the stem-and-leaf plots to find the number of tests and the range, mean, and median longevity. (All but the mean are easily determined by simple counting techniques.)

Ask students to fill in the table at the bottom of the page. (They can find the exact amounts for the number of tests, the range, and the median. They should estimate the mean.)

2. Ask students to enter the data for company A into the statistics mode on their calculator. Ask them to use their statistics function keys to fill in the mean, the standard deviation, and the number of tests for company A in the table on BLM 9. They can recount or use their results from the first table to fill in the median and range. Repeat the process for companies B and C.

Activity 2: Compare and Contrast the Statistics

1. Ask students to look at the completed chart on BLM 9 and to compare and contrast the various statistics among the companies. Students will notice that for each company, the mean number of hours (69.3333...) and the number of bulbs (30) are constant, but the range, the mode, and the median vary. Help students see how the shape of the stem-and-leaf plot changes as the range, the mode, and the median vary. Ask questions such as the following:

How is the shape of company B's stem-and-leaf plot different from that of company A? [Company A's plot is flatter and more spread out.] *Now look at the range for each company. How is the shape of the plot reflected in the range?* [The range is wider for company A, so the plot is wider.]

How is the shape of company C's stem-and-leaf plot different from that of company A? [Company A's plot seems to bulge at the top end, whereas company C's plot bulges more in the middle.] *How is the shape of the plot related to the median of the two companies?* [The median for company A will be closer to the top than the median for company B.]

2. Ask students to examine the numbers given for standard deviation. Pose the following questions:

Compare companies A and B. How do the plots compare? How do the standard deviations compare?

Compare companies A and C. Compare companies B and C. How do their plots compare? How do their standard deviations compare? [The more spread out the plot, the larger the standard deviation]

Write a statement telling what you think the standard deviation tells someone.

REASONING: This "discovery" approach to standard deviation is based on inductive reasoning. Rather than mathematically deriving the formula for standard deviation, students are given information about standard deviation and asked to make reasonable conjectures about the meaning based on the information and patterns they observe.

Evaluation. If middle school students have not had previous experience writing verbal descriptions of mathematical relationships, they may have a difficult time with generalized questions such as the one immediately above. One way to initiate the use of writing in mathematics is to pose questions about specific situations:

> Two seventh-grade students had exactly the same average spelling score for the first semester of school. Janice and Albert both averaged 90 percent on their spelling tests. However, when Janice and Albert entered their individual scores into the scientific calculator, Janice found that her standard deviation was 3.8, and Albert found that his standard deviation was 1.5. Can you describe how Janice's and Albert's performances on spelling quizzes differed throughout the semester? Explain your answer.

♦ ♦ ♦ ♦ ♦ ♦ ♦ ♦

3. To extend the concept of standard deviation as a "measure of spread," review the information about standard deviation at the bottom of BLM 9. To find the interval of one standard deviation around the mean, subtract the standard deviation from the mean and add the standard deviation to the mean. When the data are normally distributed, about two-thirds of the data will always fall inside this interval. Try this procedure with each company.

***CONNECTIONS:* Note the similarity in structure between standard deviation in which two-thirds of the data fall within one standard deviation from each side of the mean and a box plot in which 50 percent of the data falls within the box.**

Activity 3: Questions to Ponder

Use the following questions in class to help students intuitively explore the meaning of standard deviation as they look back at the data sets on BLM 9.

Which company has the largest standard deviation? [Company A] *The smallest?* [Company B]

When you look at the numbers, are the standard deviations reasonable? Predictable? [Yes, you can see by the spread of the data. You can predict which will have the largest and the smallest standard deviation by looking at the relative spreads of the data. Finding an approximate value takes some practice. First, look for the middle two thirds of the data, and then estimate the distance of the extremes from the mean.]

When the standard deviation of a group of numbers is small, ______________________________ (Complete the sentence.) [The spread or dispersion of the data is small.]

How would the stem-and-leaf plot data look different for two sets of data in which the means were different and the standard deviations were the same? [The amount of spread would be about the same, but the central tendencies would be different.]

Each company advertises the same average burning length. Are they being truthful? What don't they tell you? From which company would you want to buy bulbs? Why?

Evaluation. Teachers should furnish opportunities for students to encounter the standard deviation in varied contexts and from varied perspectives. The new settings can be used to evaluate whether or not the students' understanding is solidifying. When students can transfer and modify their knowledge to fit new situations, you have evidence that their knowledge of the topic is strong, flexible, and solid. Some suggestions for varying settings and perspectives follow:

1. Suppose that these sets of data were the measurements of the diameters of ball bearings that fit into a machine. The machine works best with bearings with a diameter of 69mm, but will tolerate a plus or minus 2mm difference. Which group would be better to use? Why?

2. Suppose that these sets of data were salaries (in thousands) of the top executives in three companies. For which company would you rather work? Why?

3. Suppose that the sets of data were three basketball teams' scores over a season. How would you expect the teams to be similar? Different? How can you tell which team probably won the most games? Which team would you rather be on? Play against? What would you predict as a score for each team in its next game?

4. What other situations could these numbers be from?

5. a. Compile a set of fifty numbers that have a mean of 40 and a standard deviation of about 5. Plan ahead and use your calculator.

b. Create another set of fifty numbers that have a mean of 40 and a standard deviation of about 10.

c. How are these sets of data alike? How are they different?

6. In what type of situation might it be good to have a small standard deviation? A large standard deviation?

Closing Comments

This activity illustrates the powerful blend of technology, problem solving, and statistics. Students are able to investigate and form intuitions about a mathematical topic before they learn the formal mathematical derivations of the standard deviation. The power gained by having students use tools allows them to gain insights into concepts and ideas that were not previously in the curriculum.

CHAPTER 4
A FOCUS ON REASONING

Quantitative reasoning will be a critical skill in the twenty-first century. Virtually all problem situations will be addressed by sorting through and analyzing great amounts of information to make decisions for future actions. The incredible capability of computers to store vast amounts of data guarantees that most decision making will be based on analyses of quantitative data. The reasoning skills underlying those decision-making processes need to be developed from an early age, while students are first learning concepts, first gathering data, and first learning to make informed decisions. This section examines three types of reasoning that can be included in the middle school experience: reasoning with quantity, reasoning about cause-and-effect, and reasoning about chance.

REASONING WITH QUANTITY

Paulos (1989, p. 66) illustrated a common medical use of percentages that is often misinterpreted. He posed a situation in which a person tests "positive" for a particular disease that occurs in 0.5 percent of the population. Given that the test has been found to be accurate 98 percent of the time, Paulos reasons through the problem with the reader to the surprising conclusion that, on receiving news of a positive test result, the person's chances of having the disease would be only around 20 percent. The following activity, adapted from Paulos, furnishes questions and diagrams to help students reason through the problem using their basic knowledge of percentages.

PROBLEM SOLVING: Problem solving in the real world does not always involve finding an "answer." More often people are faced with situations in which they must make qualitative judgments, form arguments, and make nonmathematical decisions based on quantitative information.

ILLUSTRATION 9: MEDICAL TESTS: HOW WORRIED SHOULD YOU BE?

This illustration gives students an opportunity to evaluate a situation by using their basic knowledge of percentages. They must reason through the quantitative information to make a judgment rather than to "find an answer."

Activity 1: Getting to Know the Problem

Pose this situation to students:

> A medical test for a particular disease that occurs in 1 percent of the population has been found to be 97 percent accurate. Suppose the test were run on you, and your results were positive. How concerned would you be about actually having the disease?

Students should spend some time trying to understand the problem setting by discussing such questions as these:

Out of 100 people, how many would you expect to have the disease? 200 people? 1000 people? 10 000 people?

What does it mean to have "positive" or "negative" test results?

What does it mean that the test is 97 percent accurate? [When testing people who do have the disease, the test detects the disease 97 percent of the time. When testing people who do not have the disease, the test results are correctly negative 97 percent of the time.]

When people who have the disease are tested, what percentage of the time are the results incorrectly negative? [3 percent]

When people who do not have the disease are tested, what percentage

PROBLEM SOLVING: Solving problems in the real world often means that the majority of time and effort is spent on understanding the problem, not on carrying out the plan, whereas most school experiences emphasize devising and executing a plan. Classrooms that reflect the spirit of the Curriculum and Evaluation Standards strive to balance the types of problems that students encounter in mathematics class.

of the time are the results incorrectly positive? [3 percent]

Activity 2: Evaluating the Situation

To evaluate the situation in Activity 1, assume there is a population of 10 000 people. Pose the following questions, using a diagram to help the students understand the situation (see fig. 7). On the chalkboard, clearly write and label each answer.

a. Imagine there are 10 000 people sitting in an auditorium.

How many of those people would you expect to have the disease? [100]

How many people in the auditorium would not have the disease? [There are 10 000 people in all, and 100 have the disease, so there would be 10 000 – 100 disease-free people. Nine thousand nine-hundred people would not have the disease.] Ask students to create diagrams to represent this situation. Figure 7 illustrates the situation.

Fig. 7

b. Recall that in the original situation, your test results were positive. So, let's think about the people in this auditorium who tested positive.

Of all the people who do have the disease, how many of them tested positive? [97 percent of the 100 people is 97.]

Of all the people who do not *have the disease, how many of them tested positive?* [3 percent of 9900 is 297.]

Show this on the representation.

c. Since you tested positive, consider all the people in the auditorium who tested positive.

How many people altogether tested positive? [97 + 297 = 394]

How many really have the disease? [Only 97]

What fraction/percentage of those who tested positive actually have the disease? [97/394, or about 25 percent]

So, what are your chances of actually having the disease if you tested positive? [About 25 percent]

Activity 3: Extensions

1. What do you think the physician will do next? [The doctor will run the test again.]

2. How many tests do you think the physician will want to run to be very certain that the person is sick? [The doctor will probably run the test two or three times.]

3. Ask the students to describe all possible situations that are reflected in each of the following circumstances. Then ask for implications of the situations, considering that the person being tested does not know whether or not he or she has the disease.

a. Negative test followed by positive test

Situation 1: A person does not have the disease. The first test was accurate and the second test was inaccurate—a "false positive."

Situation 2: A person does have the disease. The first test was inaccurate—a "false negative"—but the second test was accurate.

Implications: Since the first test was negative, the chances are *very* small that the person actually has the disease. It is surprising that a second test was run.

b. Positive test followed by negative test

Situation 1: A person has the disease and the first test was accurate. The second test was a "false negative."

Situation 2: A person does not have the disease, so the first test was a "false positive." The second test correctly assessed that no disease was present.

Implications: Since the first test was positive, there is some valid concern. However, since the chances of having a "false positive" are greater than having a "true positive," high levels of anxiety would not be warranted until the second test results come back. The negative results of the second test would probably be sufficient to conclude that the person does not have the disease.

c. Two positive tests, and then two negative tests

d. Negative, positive, and then negative tests

e. Positive, positive, and then negative tests

f. Positive, negative, and then positive tests

g. Other combinations of positive and negative test results

Evaluation. Small group settings afford opportunities for the teacher to hear and evaluate students' verbal reasoning skills. Pose a variety of problem settings for students to solve in small groups.

a. Suppose you tested positive for a disease that occurs in 2 percent of the population. The test used is known to be 99 percent accurate. What are the chances you actually have the disease?

b. Discuss the meaning of "false positives," "false negatives," "true positives," and "true negatives."

c. For each of the situations above, find the number of "false positives," "false negatives," "true positives," and "true negatives."

Since the goal of the illustration is to develop reasoning skills, the evaluation of students should be based on process objectives. Small group settings provide opportunities for students to reason and explain ideas aloud. The teacher can use a checklist while he or she observes and listens to students in small groups as they solve problems. An example of an evaluation checklist follows:

Group Members:____

(+) behavior observed
(-) opposing behavior observed
(n) no opportunity for observing this behavior

____Each group member is engaged in the group task.

____Students are making reasonable and understandable explanations to each other.

____Alternative explanations are consciously sought.

____All proposed ideas are seriously considered.

____Group members are making sure they all understand the information.

____Students are monitoring the accuracy of their information.

____Students "look back" to monitor their present solution attempt.

Closing Comments

The goal of this illustration is to furnish a problem setting for developing students' reasoning processes. Unlike instruction aimed at building competence with skills, the instruction required for this illustration needs to be based on questioning techniques that help students develop flexibility in thinking, analytic consideration of alternatives, and reflection on their own thinking processes.

REASONING ABOUT CAUSE AND EFFECT

Children and adults learn much about the world around them by noticing when one event commonly follows another. For example, a child notices that every time his parent turns the sink faucet handle, water comes out. So, when the child wants water to come out of the faucet, he turns the handle. He has learned that turning the handle causes water to come out.

Sometimes cause-and-effect perceptions are oversimplified. A baby's cries are often quickly followed by the presentation of a warm bottle of milk. The baby implicitly learns that crying causes nutrition to come. Although the baby certainly is not capable of understanding the complex chain of causal events that the crying evokes, the use of the oversimplified cause-and-effect relationship does, in fact, work. When the baby cries, food generally comes.

Children also believe in a cause-and-effect relationship where none exists. Imagine a child who is frightened of the dark because of an imagined boogie man that lives under the bed. The child develops rituals (for example, tightly tucking in the covers and falling asleep under the pillow) that are believed to prevent the boogie man from attacking. Clearly, the ritual does not cause the nonexistent boogie man to stay away, but the ritual appears to work for the child. So, the child strongly believes that the ritual does indeed cause the boogie man to stay away.

Adults, too, are not immune to making invalid cause-and-effect connections. For example, a person may believe that taking high doses of vitamin C at the onset of a cold will cure the cold in five to seven days. In reality, a cold runs its course in the same time period and would have cleared up in that amount of time anyway. Therefore, although the vitamin C treatment did not necessarily cure the cold, the person incorrectly believes it did.

Using data and information to make predictions and decisions in problem situations requires sound reasoning about causal relationships. Numerical data may show that two events do in fact correspond (i.e., have a high degree of correlation), but the relationship is not necessarily causal. Reasoning must be used to develop a sound explanation for the relationship, to rule out alternative explanations, and to make a logical case for cause and effect.

The following classroom illustration uses varied paired events to help students begin to assess critically whether two events have a causal relationship. Few middle school students have had experiences where they have critically assessed whether two coinciding events are connected causally, yet opportunities for developing such reasoning occur repeatedly in everyday applied settings. Explicit attention to developing reasoning processes that help students decide whether causal relationships actually exist is needed in the middle school data and chance curriculum.

ILLUSTRATION 10: COINCIDENCE, SUPERSTITION, OR FACT?

Eunice D. Goldberg

Three newspaper articles are used for a discussion that focuses on the reasoning processes needed to decide whether one event certainly causes another. The first activity addresses an absurd pairing of events that clearly have no relationship. The second activity addresses a superstition where the possibility of a causal connection may be plausible to the superstitious. The third activity considers a pair of events that are highly correlated. This situation even has an explanation that could justify a causal connection, yet there is reasonable doubt about the relationship. The illustration ties the three examples together at the end by helping students evaluate causal relationships by questioning and searching for alternative explanations.

Materials

blackline masters 10 and 11

Preparation

Reproduce classroom sets of BLM 10 and 11.

Activity 1: Pizza and Subway Fares

Pass out the article "If You Understand Pizza, You Understand Subway

Fares" (*New York Times*, Dec. 1985) on BLM 10. This humorous editorial ponders the fact that increases in subway fares seem to occur whenever the cost of a slice of pizza is raised. Since this article presents two coinciding events for which a causal link is obviously absurd, it makes a good introduction to situations where cause and effect cannot necessarily be inferred from the timing of occurrences. Pose the following questions for discussion:

Are coincidental occurrences (two events occurring around the same time) evidence of cause? Do we have the right to say that events that occur in close proximity cause each other? For example, if every time a baby is born someone gets a flat tire, is it correct to assign the cause of a flat tire to the birth of a baby?

What might be "reasonable" causes of a fare increase on the subway? Are transit prices (or any prices) always related to underlying expenses? Might there be other considerations, such as what the market will bear? Who determines this? Think of two identical pieces of clothing—one with a designer label, one without—and the variation in prices.

Activity 2: "Dear Abby"

Read the "Dear Abby" article (Universal Press Syndicate 1985) on BLM 11, which addresses the role of superstition in making predictions about future events. This letter makes two incorrectly reasoned causal links: (1) the person who catches a wedding bouquet will be the next person to wed; (2) the divorce of the original couple nullifies the effects of catching the bouquet.

Pose the following questions for discussion:

Are unrelated events being used to predict the future? What are superstitions? Can you think of other superstitions that incorrectly imply cause-and-effect relationships between two events? [e.g., Don't step on a crack or you will break your mother's back. Don't let a black cat cross your path or you will have bad luck.]

Is it possible that brides do certain things to increase the probability that the person who catches the bouquet will be the next to wed? Consider these examples: (1) Does the bride limit the group she tosses to? [Yes, usually to single, eligible females] (2) Does the bride aim the bouquet at a person who is engaged to be married in the near future? [Very often this is the case.]

Discuss the possible outcomes if the bride were to toss her bouquet randomly to any woman in the room. What if her mother (who, we'll assume, is happily married to her father) caught it? Does that mean she'll get divorced and be remarried before anyone else in the room? What if her eight-year-old cousin caught it?

Activity 3: Bad News for Cola-quaffing Athletes

Read the "Cola-quaffing Athletes" (Scripps Howard News Service, 1980) article on BLM 11, which discusses a possible causal relationship between the amount of carbonated beverages consumed by female athletes and the frailty of their bones, especially after age forty. In this situation, the quantitative information clearly indicates a numerical correlation, but causation cannot readily be assumed or ruled out. Typical of many situations of data analysis, reasoning about cause and effect must go beyond the quantitative information and rule out alternative explanations. Two major issues need to be considered:

What constitutes sufficient proof of cause?

◆ ◆ ◆ ◆ ◆ ◆ ◆ ◆

What other explanations exist for the events that occurred?

After reading the article on blackline master 11, begin the following discussions:

- Discuss how, unlike the events in the other articles, there may be an explainable relationship among the events that occurred—athletic participation, consumption of a quantity of carbonated beverages, and incidence of broken bones—in the same person.
- Discuss whether the explanation given—the interference of phosphoric acid with calcium absorption—seems reasonable. Since the specific science concepts may not be readily known by the students, the science teacher can become involved in evaluating the plausibility of this explanation.
- Discuss the method of assessing cause by comparing two or more groups on the variables of interest. In this case, the variable of interest is the incidence of bone fractures after age forty. Have the class discuss the ways the groups could be split to find some relationships: women who exercise versus women who do not exercise; cola drinkers versus non–cola drinkers; cola drinkers who exercise versus cola drinkers who do not exercise; non–cola drinkers who exercise versus non–cola drinkers who do not exercise; and so on. Have the students explain what kind of information they would obtain from each comparison.
- Discuss the importance of carefully defining terms. Issues such as deciding what constitutes a cola drinker or a person who exercises need to be decided. There is a difference between a person who drinks one cola every few days and a person who drinks several colas every day. The same is true for exercise. Is a person who plays tennis only during the summer to be considered the same as a person who jogs several times a week? The answers to these questions are decided by the questions to be answered by the study.
- Alternative explanations for the findings need to be explored. The article mentions the possibility that bone frailty may be the result of reduced milk consumption by women. Have the class discuss other possibilities. For example, is it possible that more athletes break their bones because their activities present them with more opportunities to do so?

CONNECTIONS AND PROBLEM SOLVING: Real-world problems very often have insufficient data, yet the problem solver must solve the problem or make a decision. If possible, the problem solver makes connections to outside resources, human experts, and other disciplines to obtain the needed information. Often, however, the problem solver must make a decision before all pertinent information is gathered.

Activity 4: Looking Back and Extensions

1. Give the following written assignment: Write about possible causes for the outcomes indicated in the articles. Explain why your listed possible causes are reasonable.

2. Have the students find other newspaper articles that attribute cause to events in a questionable manner.

3. Give students a list of pairs of events—some that are clearly unrelated and some that are not. Have students identify those events that are clearly unrelated and explain their decision for that choice. For events that are not as easily classified, have students search for alternative reasons for the coincidental occurrence and make a case for why the two may or may not be causally related.

Evaluation. One way to evaluate students' understanding of cause-and-effect relationships is to give students pairs of events. The students are to indicate whether they are certain or uncertain of the causal relationship and to include a brief explanation of their classification. Responses should reflect three considerations. First, the first event must *always* be followed by the second event. Second, there must be some logical explanation that connects the two. Third, any other plausible explanations must be ruled out.

Example 1: Every time the refrigerator door is opened, the light bulb inside goes on. Therefore, opening and closing the door causes the light to go on and off.

[Possible response: "This cause-and-effect relationship is certain because a switch connected to the electrical current of the light bulb is pushed and released as the door is opened and closed. So, opening and closing the refrigerator door causes the light to go on and off." This relationship is so commonly known that a student may not feel compelled to examine the "A always follows B" relationship and to consider other plausible explanations. Students who address the fact that the light bulb always seems to be on whenever they look are really being more thoughtful. They may make a valid conjecture that the light bulb is on at all times.]

Example 2: Everytime I forget my umbrella, it rains on my way to school. So, forgetting my umbrella causes rain.

[Possible response: "I think this is a certain cause-and-effect relationship because it *does* seem that every time I forget my umbrella, it does rain." This student's reasoning is faulty. First, he is ignoring or forgetting about the many nice days when the umbrella has been forgotten and it is sunny and does not rain. So, the student is violating the "A always follows B" relationship. Further, the student gives no explanation for the causal connection.]

Example 3: Every spring, the ocean salmon travel back to their places of birth to spawn. The temperature of the air must cause them to return home.

CONNECTIONS: Determining whether there is a cause-and-effect relationship between two events requires that a high numerical correlation exist between two events, that a logical explanation for the cause-and-effect connection be made, and that alternative explanations be systematically ruled out through logical reasoning. Students' numerical understandings must interact with contextual understanding of the problem situation to reason about cause-and-effect relationships.

[Possible response: "Although the temperature does consistently change seasonally and the salmon also return seasonally, there are other cyclic things that may cause the salmon to return home. It could be some genetic biological clock, the position of the sun, or even the position of the stars that the salmon respond to." This explanation addresses all three aspects clearly and specifically.]

Closing Comments

When bombarded with numbers that indicate high levels of coincidence between two events, many people often assume that the numbers "prove" the casual relationship. However, that is not the case. Causality must be based on logical explanations that link the two events, *and* plausible alternatives must be ruled out. For example, the announcement by the United States Surgeon General that smoking *is* hazardous to one's health was based not only on numerical information about the high coincidence of smoking and lung cancer but also on biological explanations and systematic consideration of other possible causes. Lobbyists for the tobacco industry do not argue about the high numerical correspondence, but rather, that factors other than cigarette smoking contribute significantly to the high incidence of lung cancer. Students should be skeptical of implied cause-and-effect relationships based only on numbers. They also need to consider logical explanations and to systematically consider and rule out alternative explanations before they conclude that a cause-and-effect relationship exists.

♦ ♦ ♦ ♦ ♦ ♦ ♦ ♦

REASONING ABOUT CHANCE

Four types of probability have been described by Hawkins and Kapedia (1984). *A priori probability* is "the probability obtained by making an assumption of equal likelihood in the same space" (p. 349). When students are asked to identify the probability of rolling a 5 on a number cube by simply examining the number cube, they are being asked to determine the a priori (or theoretical) probability. *Frequentist probability* is "calculated from observed relative frequencies of different outcomes in repeated trials" (p. 349). This is illustrated when students actually roll a number cube 100 times and find the "experimental" or "empirical" probability of rolling a 5 by comparing the actual number of 5s rolled to the 100 rolls. *Subjective* or *intuitive probabilities* are "to a greater or lesser extent . . . an expression of personal belief or perception" (p. 349). For example, most card players believe that the chance of being dealt a royal flush in a poker hand is very small, even though they have never formally calculated the probability or systematically collected and analyzed data about the event. Sometimes people develop incorrect subjective probability beliefs such as the Gambler's Fallacy (i.e., following a long run of rolling no sums of seven on two dice, the probability of rolling a seven on the subsequent roll is increased). *Formal probability,* in contrast, "is calculated precisely using the mathematical laws of probability" (p. 349). Many mathematics teachers remember calculating in their college probability and statistics class the probability of being dealt a royal flush in poker.

For middle school students to learn to reason logically about the many forms of chance, they must build intuitions from real experiences. These intuitions will help the students develop formal reasoning skills for using mathematical analysis of probability. The following activity begins with a game in which students acquire an intuitive feeling about the chances of "winning" and concludes with a formal analysis of the various strategies and plays. Students' ability to reason about chance is grounded in experiences that are then linked to formal analyses.

ILLUSTRATION 11: MONTANA RED DOG

Arthur Hyde

Montana red dog, a card game from the Old West, furnishes students with an opportunity to work in small groups collecting data, making predictions, and determining probabilities. The essence of the game involves having students predict whether or not one of the cards from their four-card hand can beat a card that the teacher turns up. To gain points, the students' cards must be higher *and of the same suit* as the teacher's card. Students become increasingly sophisticated in their probabilistic reasoning as they develop strategies to enhance the accuracy of their predictions.

Students should be arranged in groups of two or three to make *ten* groups. The activity works best with an oversized deck of standard playing cards, which can be purchased at magic shops or through catalogs. The jokers are removed, and the cards in each suit go from a low of two to a high of ace. In each game, the teacher shuffles and deals four cards face down to each group. (Thus, forty cards are dealt and twelve cards remain with the teacher.) Groups may look at their own cards but not at each others'. The teacher's cards must not be shown to the class. Each group will get a turn at trying to beat a different card from the teacher's pile.

Materials

copies of figure 8
oversized deck of playing cards

Activity 1: Understanding the Game

Game 1: After shuffling the cards, deal out four cards face down to each of the ten groups. Group members should examine their cards privately and decide how confident they are in beating the top card of the teacher's deck. They should come to consensus on one of the following numerical levels of confidence (i.e., confidence points):

(0) We don't think we can beat it.
(1) We think we might beat it.
(2) We are pretty sure we can beat it.
(3) We are certain we can beat it.

One by one, each group announces its confidence points prior to the display of the top card of the teacher's deck. The group then displays its cards. If one of the cards beats the teacher's, the class scores the announced number of confidence points. If not, the teacher scores the same number of points. Play continues until all groups have had a turn.

Game 2: Play another practice game with the students, but this time provide some guidance for making better judgments for confidence points.

1. Deal out four cards face down to each of the ten groups.

2. Pass out multiple copies of figure 8, which has four columns headed with each suit and with ace through two in descending order below. Using one copy of figure 8, each group should draw circles around the four cards that they have been dealt.

♠ SPADES	♥ HEARTS	♦ DIAMONDS	♣ CLUBS
A	A	A	A
K	K	K	K
Q	Q	Q	Q
J	J	J	J
10	10	10	10
9	9	9	9
8	8	8	8
7	7	7	7
6	6	6	6
5	5	5	5
4	4	4	4
3	3	3	3
2	2	2	2

Fig. 8

3. Students should determine which cards they can definitely beat in each suit (below the circled cards) and write the number of cards below the column for that suit. (See fig. 9 for an example hand.) You may need to remind them that if they have no cards in a suit, the number that should be written below that column is 0. Students should add the four

bottom column numbers and write this sum in the lower right-hand corner of the page. This sum is the CAN BEAT number.

4. Next, students should determine the number of cards that they cannot beat—all cards above the highest circled cards in any suit and *all* thirteen cards in a suit that has no circled numbers. They should write these four numbers above the columns near the names of the suits, add them, and write this sum in the upper right-hand corner. This sum is the CANNOT BEAT number. A simple check is that the two sums must total 48 (the total number of cards in the deck minus their four cards).

♠ SPADES	♥ HEARTS	♦ DIAMONDS	♣ CLUBS	
4	13	3	7	(27) (Cannot beat)
A	A	A	A	
K	K	K	K	
Q	Q	Q	Q	
J	J	(J)	J	
(10)	10	10	10	
9	9	9	9	
8	8	8	8	
7	7	7	(7)	
6	6	6	6	
(5)	5	5	5	
4	4	4	4	
3	3	3	3	
2	2	2	2	
7	0	9	5	(21) (Can beat)

Fig. 9

5. Use one group of students to illustrate how the information recorded on its copy of the figure can be used to decide on confidence points. The selected group should show its four cards and state its CAN BEAT and CANNOT BEAT numbers and its confidence points. The teacher can record these in a chart. The chart in figure 10 uses the information from figure 9. (During a game, make such a chart for later analysis and reflection.) Then the teacher's card is shown and the outcome (W or L) is recorded along with the score.

Group	CAN BEAT	CANNOT BEAT	Total	Confidence Level	Outcome
xxx	21	27	48	(2)	(W or L)

Fig. 10

6. The full game should now continue with the rest of the groups, and postgame discussion should be encouraged by posing the following questions:

Who had the best situation for holding the winning card, the teacher or the class? How can you tell?

Who had the worst situation? Explain.

How should the CAN BEAT and CANNOT BEAT numbers affect the number of confidence points chosen?

Are you likely to make higher confidence ratings at the beginning of the game or near the end? Explain your answer.

Activity 2: Playing the Game

1. Shuffle the deck and deal out four cards to each of the ten groups. Each group should circle its own four cards on a clean copy of figure 8, keeping its cards and record sheet private. The first group should tell its CAN BEAT and CANNOT BEAT numbers, declare its confidence points, and, following the display of the teacher's card, show its cards to the class. Record the information in a chart for a postgame discussion, and assign the appropriate number of points to the teacher or the class.

2. As the game continues, the groups will be able to cross off the cards used during the game. As each group exposes its cards, the other groups can recalculate their CAN BEAT and CANNOT BEAT numbers. The game continues in this fashion, with each successive group having five fewer cards to consider in its chances. The students in the tenth group will have the benefit of knowing all the cards except for the three remaining in the teacher's pile.

3. Postgame discussion questions can revolve around each group's selected confidence points and its chances of winning at different points in the game. For example:

USING TECHNOLOGY: Using calculators to determine and compare probabilities frees the students from tedious computations and allows them to focus on thoughtful processes as they consider how to respond to the teacher's challenging questions.

Team 1 (first round): 39 CAN BEATS and 4 CANNOT BEATS
Team 3 (third round): 25 CAN BEATS and 3 CANNOT BEATS

Which of the above teams has a better chance of beating the teacher's next card? Explain. [You can directly calculate the probability of each situation and compare the data. Team 1 has a 39 out of 43 (39/43) chance of beating the teacher's cards. That is a 90.7 percent chance of beating the teacher. The probability that Team 3 will beat the teacher is 25/28, or 89.3 percent.]

Is it likely that the teams above will have different confidence points? Explain. [Since the chances of beating the teacher are nearly the same, the confidence points should be about the same.]

What situations would contribute to groups selecting high confidence points? Low confidence points?

Is it possible to have a 90 percent chance of beating the teacher and still lose points? Describe such a situation.

Is it possible to have a 100 percent chance of beating the teacher and still lose points? Explain.

Is it possible to have a 10 percent chance of beating the teacher and still gain three points? Describe such a situation.

Is it possible to have a 0 percent chance of beating the teacher and still gain three points? Explain.

PROBLEM SOLVING: One way to help students acquire the problem-solving strategy "think of a similar problem" is to have students develop the similar problem themselves.

Activity 3: Extension

Ask student groups to modify the rules of the game to fit the following situations:

a. The student groups would be playing against one another, with the teacher as recorder and card-turner.

b. Two individuals could sit down and play the game.

c. Any number of individuals could play the game.

♦ ♦ ♦ ♦ ♦ ♦ ♦ ♦

Closing Comments

This game affords an opportunity for students to measure intuitively their own level of confidence in uncertain situations, to see how mathematically determined probabilities can help them better choose their levels of confidence, and to compare and contrast the mathematically expected outcomes with the actual outcomes. Since this game nearly always involves situations of *uncertainty,* some groups will lose points despite high levels of probability or will win even with low levels of probability. With proper discussion, these surprises can also enrich students' conceptual understandings of, and reasoning with, probability.

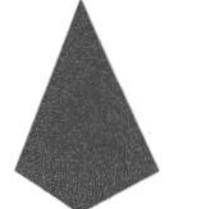

CHAPTER 5
A FOCUS ON CONNECTIONS

Throughout this book, the data and probability explorations are consistently tied to other content areas, both within and beyond the field of mathematics. Meaningful analyses of data and chance occur only in context, making the study of statistics and probability a "natural" for intra- and interdisciplinary connections. The first illustration in this section involves making connections within the field of mathematics; the second illustration makes connections to language arts.

MAKING CONNECTIONS WITHIN MATHEMATICS

Classroom explorations of data and chance need not be limited to a unit on probability and statistics. For example, Illustration 12: *Counting Out* πr^2 shows how data collection, representation, and analysis can be used to investigate the relationship between the radius and the area of a circle.

ILLUSTRATION 12: COUNTING OUT πr^2

Adapted from a TIMS (Teaching Integrated Mathematics and Science) Project activity (Goldberg and Wagreich 1989)

In this activity, students study the relationship between the radius of a circle and its area. Students measure the radius of a circle, then count and estimate the number of square centimeters needed to cover the area. By using a variety of circles and obtaining a variety of measures from different classmates, students are involved in gathering, organizing, summarizing, representing, and interpreting data. When the data are represented on a coordinate graph, students face the issue of how to use the data to make the best possible predictions. Difficulties of predicting from curves lead to exploring the relationship between the area and the radius squared. This relationship turns out to be linear through (0, 0) and is, therefore, a proportional relationship, which promotes ease in making predictions.

Materials

centimeter grid paper
a variety of tin cans (small, medium, large)
a compass for each group
a centimeter ruler for each group
a calculator for each group

Preparation

Ask each student to bring one or two tin cans to school.

Activity 1. Gathering the Group Data

1. Give each group of three students multiple copies of centimeter grid paper, at least three or four tin cans in varied sizes, a centimeter ruler, a compass, and a calculator.

2. Ask students to locate circles on the tin cans. Once they have noted that the top and bottom bases of the cans form circles, briefly review the meaning of radius, diameter, and area.

◆ ◆ ◆ ◆ ◆ ◆ ◆ ◆

3. Each student should set up a table to record the measure of the radius and the area for the circles (fig. 11).

Name________________________________

Circle from	Radius	Area
Can A		
Can B		
Can C		

Fig. 11

4. To find the radius of the bottom of a can, help students locate the diameter, which is the greatest chord (or longest distance) across the circle. Measure the diameter to a tenth of a centimeter, using the centimeter ruler. Since the diameter is twice the radius, divide the measure by 2 and round to two significant digits. Record the result in the chart.

5. Determine the area of the circle in square centimeters by outlining the base of the can on centimeter grid paper. When measuring area, students are to *count* squares and piece together parts of squares to estimate whole squares. In figure 12, the two pieces on the right labeled "2" can be put together to make the second square centimeter. The top left piece and the bottom middle piece can be combined to make the third square centimeter. Continuing in this manner, the students would find an estimate of 5.5 square centimeters to be reasonable.

CONNECTIONS:

Measurements in scientific experiments are rounded to a common number of significant digits, unlike the usual practice in mathematics class of rounding to a common place value. Students should address both types of rounding techniques in mathematics class.

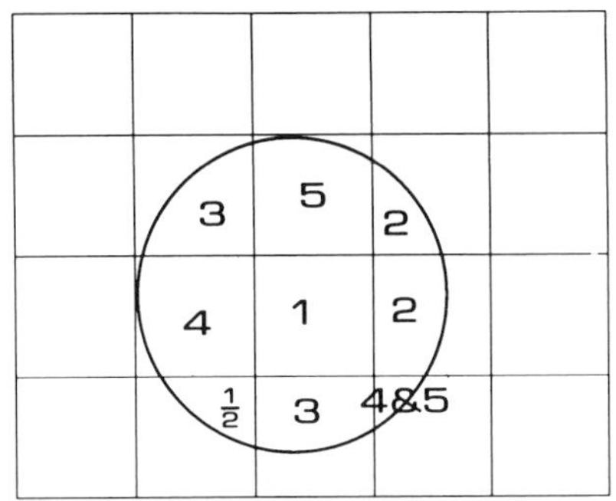

Fig. 12

6. Once each student has completed his or her individual measurements of three or four cans, the group should come to consensus about what measures to report to the class. They will note that their measures for identical cans are different. Stress that it is normal to have some variation in measurement, because all measurement is approximate. To minimize the measurement error, the groups should examine each set of measures, identify the median value, and use it in the group table. Using the median essentially eliminates all "highs" and "lows"; the resulting middle value is a good representation of the whole set of data.

CONNECTIONS: Statistics is most frequently used to minimize and deal with error. One way to deal with error in measurement is to take multiple measures of one object and then average the measures in some way. This middle value (i.e., median) is likely to be very close to the actual measure.

Example: The group members each measured the radius of can A. The results were 1.2 cm, 1.3 cm, and 1.5 cm. They should report the middle measure (1.3 cm) for further activities. Figure 13 is a sample table that one group of students formed.

Circle from	Radius	Area
Can A	1.3 cm	5.5 cm^2
Can B	2.5 cm	20 cm^2
Can C	4.0 cm	50 cm^2

Fig. 13

Evaluation. Some process skills in mathematics, such as reading a ruler, depend on hand-eye coordination and manual dexterity. As individual students are actively measuring objects, the teacher can circulate and take note of their skills. However, a more powerful assessment takes place when the results are reported to the group. Peer- and self-evaluation occur naturally when the measurements are reported by individuals and one "middle" value must be decided on. "Outliers," caused by mistakes in measuring, are immediately noticed by all group members. Often the student who reported the outlier will want to measure the object again to see what happened, simultaneously seeking confirmation and help from fellow group members. So the natural peer- and self-evaluation processes result in diagnosis and remediation within the group without teacher intervention.

Activity 2: Summarizing and Graphing the Data

1. Display each group's data in a large table on the chalkboard. If some groups measured the same objects, discuss why their measurements may have varied. Again, stress that it is normal to have some variation in measurement, because all measurement is approximate.

2. Each student should set up a graph on centimeter grid paper, putting the radius on the horizontal axis and the area on the vertical axis. The result will be a scatter plot similar to the one shown in figure 14. Students should note the curved shape of the scatter plot, especially after discussing whether or not they would expect (0, 0) to be included in the graph.

TECHNOLOGY: Graphing software capable of making scatter plots is very helpful for interpreting data from scientific experiments.

3. After students have examined the scatter plot, ask them to consider how they might draw the curve that they think best fits the data. Before they sketch the curve, ask, "Do you think the curved line should go through (0, 0)? Explain." It is intuitively appealing to students that as the radius approaches a measure of 0 cm, the area will approach 0 cm^2. Now, each student should draw in his or her idea for a "curve of best fit."

4. Ask students to use their curve of best fit to predict the examples below. After making predictions from the graph, they should make the circle with a compass on centimeter grid paper, estimate the area by counting the squares, and then compare the estimate to the prediction made from the graph. (This process may cause some students to want to adjust their curve of best fit. This is appropriate and expected.)

Example: Predict the area of a circle that has a radius of 2 cm. (See fig. 14 for making the prediction.)

Example: Predict the radius of a circle that has an area of 28 cm^2.
(See fig. 14 for making the prediction.)

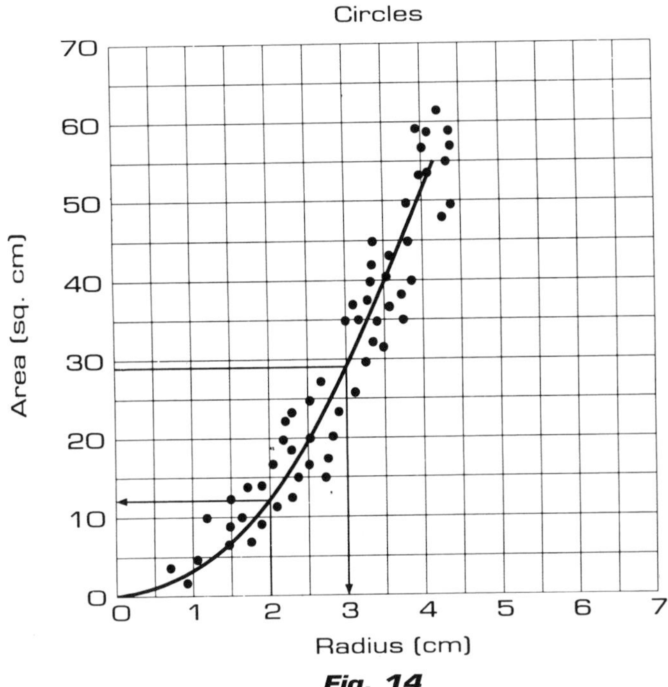

Fig. 14

Activity 3: Extending the Graph

1. Ask students to extend their curve of best fit beyond the data points. Then, students should compare their projections. Note that although most students' original curves of best fit were similar, the extensions of those curves will vary much more. (See fig. 15.)

2. Ask students to suggest ways in which they can check the accuracy of their extensions. One way is to measure the radius of a larger circle and predict the area from the graph. Then students should estimate its actual area by counting squares and should see how the estimate matches the prediction produced by using their curve of best fit. Again, they may want to adjust their extension after trying a few examples.

3. Stress the difficulty of extending the curve accurately, especially when compared to extending and predicting from a straight line. Activity 4 is designed to address this issue by straightening out the curve.

Activity 4: Straightening Out the Curve

1. Remind students of experiences in which they have made predictions from data that have had a straight line relationship. Note that extending the graph and making estimates were much easier on the straight line graph than on the curved line graph. Suggest to students that one way to straighten out this graph is to "transform" the data.

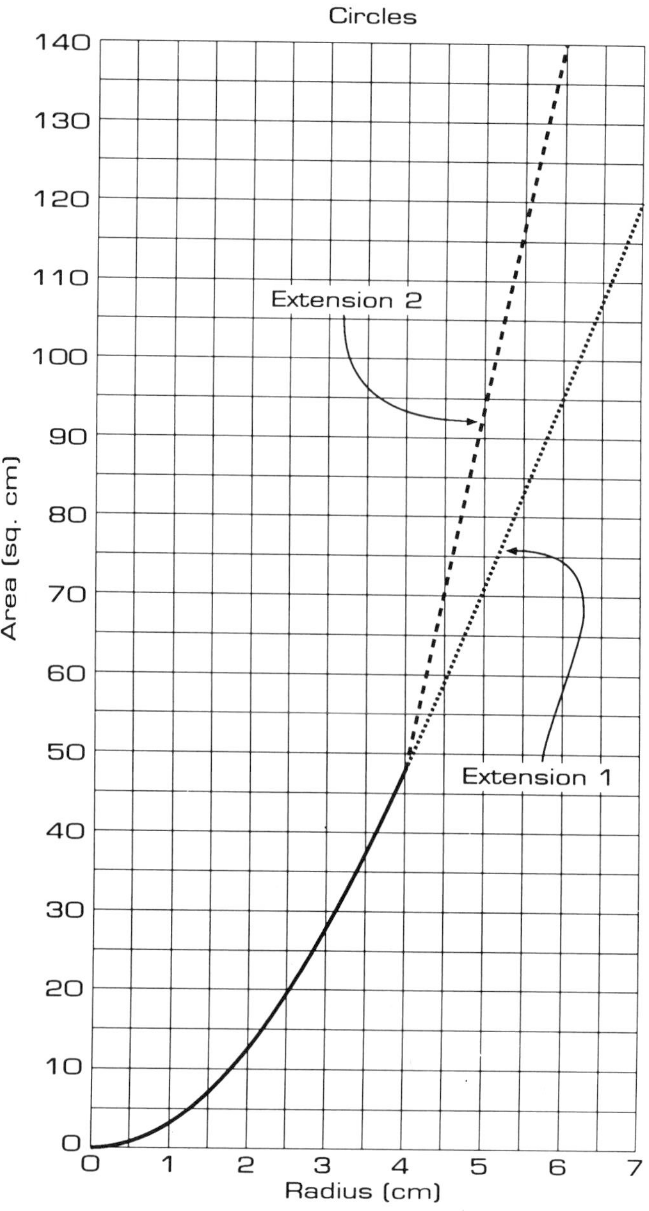

Fig. 15

2. Ask students to use their calculators to include the radius squared in a new table. (See fig. 16.)

Circle	Radius (in cm)	Radius (squared)	Area (in cm^2)
small	1.3	1.7	5.5
medium	2.5	6.25	20
large	4.0	16	50

Fig. 16

♦ ♦ ♦ ♦ ♦ ♦ ♦ ♦

3. Each group should report its radius squared and area to the class for inclusion in a class table.

4. Students should now make a graph of all the values reported by the groups, using the squared radius on the horizontal axis and the area on the vertical axis.

5. Ask students if (0, 0) should be included in the graph. [Yes] Now, ask students to describe their perception of the curve of best fit. It should now look like a straight line. Ask students to draw in what they think would be the "line of best fit" to represent the data. (See fig. 17.)

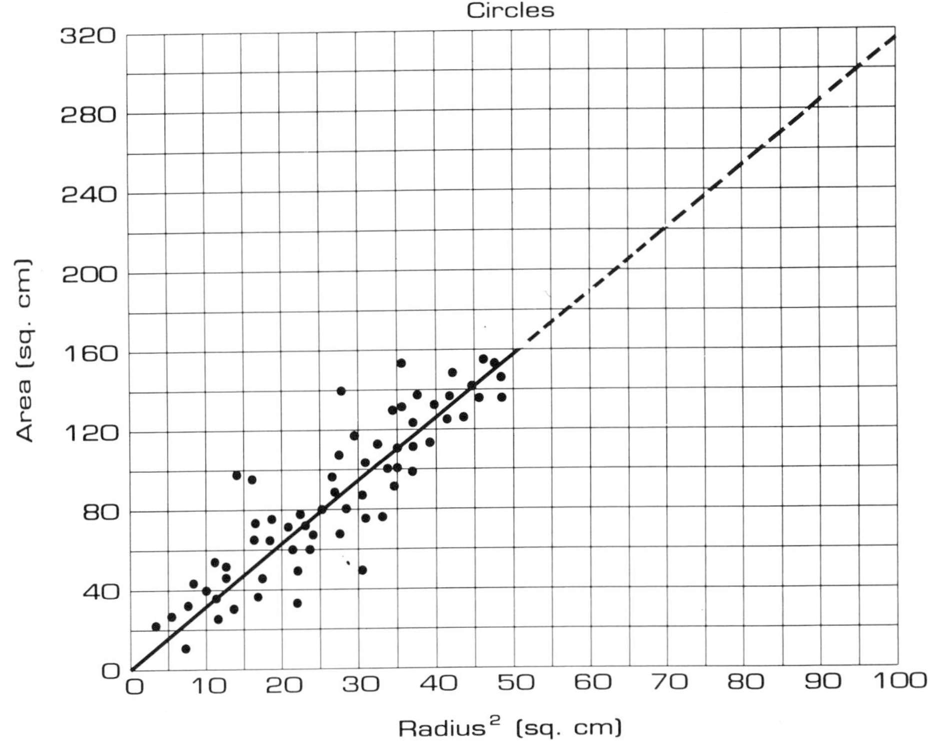

Fig. 17

6. Ask students to extend their line of best fit beyond the data points. Compare their extensions to those of their neighbors. Since the line is straight, the extensions should be very similar.

Activity 5: Interpreting the Graph

1. Ask students to work in groups to find ways in which they can use their new line of best fit to predict the area of a circle with a radius of 3 cm. Give groups about five minutes to come up with a method. Ask representatives of each group to report to the class. (Think: 3 squared is 9, so locate 9 on the horizontal axis. Find the corresponding point on the data line and follow it over to the vertical axis, as in figure 18. The area looks to be about 30 square cm.)

REASONING: Making predictions from complex graphs requires a high level of reasoning. Introducing students to complex graphs by using their own data is one way to help them more readily make predictions, since they are intimately familiar with the meaning of the data.

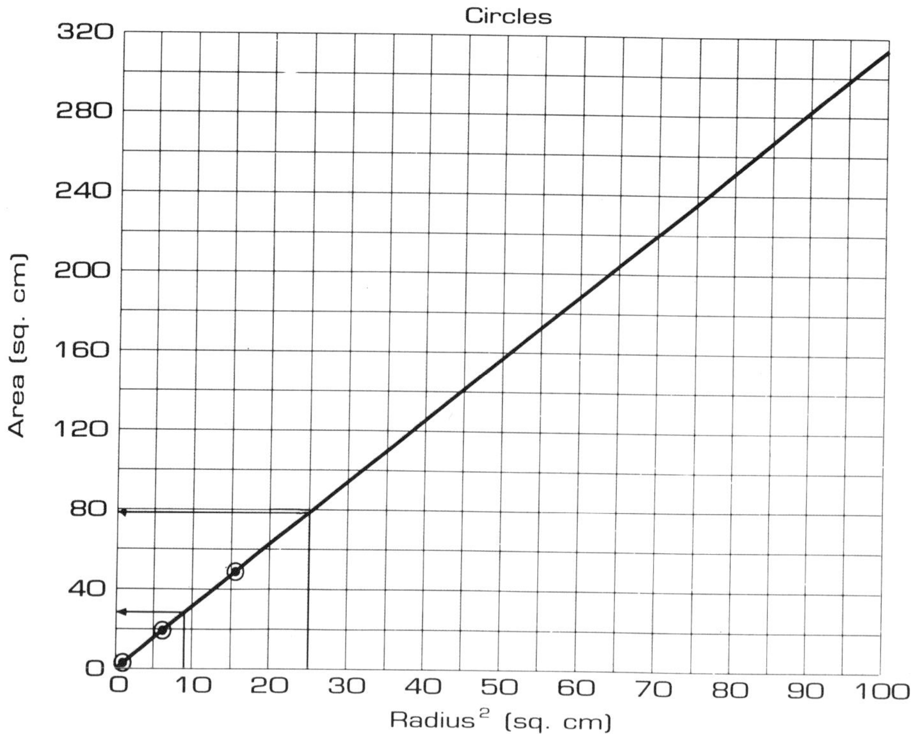

Fig. 18

2. Ask each student to use his or her line of best fit to estimate the area of circles having radii of 5 cm, 6 cm, and 10 cm.

Activity 6: Generalizing the Relationship

As a full-period class activity, ask students to work in small groups to find varied ways to predict the area of circles with radii such as these:

a. 15 cm
b. 25 cm
c. 40 cm
d. 100 cm
e. 500 cm

These examples are intended to be too big to predict directly from the graph and too cumbersome to estimate by counting squares. The goal is for students to look for a numerical relationship between the area and the radius squared. As small groups of students search for solution strategies, suggest that they set up tables and look for regularities in the number relationships. Eventually, they will note that there seems to be a proportional relationship of about 3 to 1 when comparing the area and the radius squared. A proportional relationship such as this one is characteristic of all linear graphs going through (0, 0).

TECHNOLOGY: Spreadsheet software can be a very effective tool in exploring numerical relationships between variables in large sets of data. Conjectures can be tested on all data at once and checked against the rest of the data instantaneously.

Closing Comments

This illustration was adapted from an integrated mathematics and science experiment written by Howard Goldberg and Philip Wagreich as part of the TIMS Project (1989). Students use data collection and analysis, rather than a more typical formulaic approach, to examine a standard middle school mathematics topic. The result is that students

◆ ◆ ◆ ◆ ◆ ◆ ◆ ◆

see a dynamic relationship between the area and the squared radius of a circle, rather than a static formula to be memorized. This illustration is particularly powerful in exemplifying how a traditional mathematics topic can be taught in a way that is consistent with the spirit of the *Curriculum and Evaluation Standards for School Mathematics* (NCTM 1989).

Further dynamic connections within the field of mathematics are found in Lappan et al. (1987) and Phillips et al. (1986), where geometry, fractions, and probability are interrelated.

CONNECTIONS: There are many relationships that can be explored. For example, students can explore the relationship of π to the slope of the "line of best fit" and the constant in proportional relationships.

MAKING INTERDISCIPLINARY CONNECTIONS

Treatments of data and chance are often found in the middle school science and social studies curricula. Science typically involves extensive data collection, representation and analysis, analysis of chance, deductive and inductive reasoning, statistical inference, and so on. More recently, social studies educators have been emphasizing quantitative thinking, especially in the area of data analysis, which has been called "social mathematics" by Hartoonian (1989, p. 51):

> Social mathematics includes abilities that are used when we measure or quantify social phenomena in any situation and communicate these measures to others, plus those related abilities that we need when judging the information presented to us as we decide whom to vote for, what car to purchase, or what personal economic course to follow.

Less familiar are the connections between statistics and literature and language arts. At first glance, the coupling of probability and statistics with literature and writing may seem strange. However, literature has numerous examples of probability and statistics (Boruch and Zawojewski 1987; Smeeton and Smeeton 1985; Kruskal 1978). These occurrences furnish opportunities to insert short lessons that enhance students' understanding of statistical ideas. The following illustration shows how a short story, "The Lady or the Tiger" by Frank Stockton (1980, pp. 19–24), can be used to explore data and chance as literary aspects are studied.

ILLUSTRATION 13: THE LADY OR THE TIGER

Judith S. Zawojewski and Jeri Nowakowski

This illustration provides opportunities to investigate certain and uncertain events, to assign numerical values to levels of certainty, to consider conditions that affect levels of certainty, to build the language of conditional probability, and to rank events according to their likelihood.

This short story is based on the justice system of a semibarbaric king. His trial system involves putting a suspect in an open, public arena with two doors; behind one door is a beautiful lady ready for marriage, and behind the other door is a ferocious tiger. The suspect selects one door, and his innocence or guilt is determined immediately—the suspect is either married or mauled. The plot of the story involves a conflict between the king and his daughter over her handsome lover, who is of common bloodline and is considered unworthy in the king's eyes. As the king plans to put the young man on trial for treason, the princess ponders what to do with her knowledge of the fate behind each door. The author makes it clear that although the princess does not want her lover to be devoured by a tiger, she also is semibarbaric and very jealous. The princess does not want to see her lover marry another woman. The story ends as the young man looks to the princess for guidance on which door to choose, and the reader is left to imagine an ending to the story.

Materials

"The Lady or the Tiger," a short story by Frank Stockton (1980)

Preparation

Have students read the short story before class.

Activity 1: Identifying Certain and Uncertain Events

1. Have the students identify everyday events that are certain and uncertain. For example, the rising and the setting of the sun are certain events. The scheduled tides of the ocean are certain. The election of a specific politician to office is uncertain. The continuing success of a business is uncertain. This discussion should bring out the idea of varying levels of uncertainty. For example, although the continued success of the Amoco corporation is *almost* certain (at least in the short run), the likelihood of success of a brand new dry cleaning business is much more uncertain.

2. With the students, make a list of certain and uncertain events from the story, and give reasons for each classification. For example, it is certain that the princess and her lover will not marry, because if he opens the tiger's door he will surely die, and if he opens the beautiful woman's door, he will marry this other woman. However, *is* it certain that they won't marry? Is it possible for the princess to make a public plea to her father and get him to relent?

Activity 2: Identifying Influencing Conditions

1. Use some of the uncertain events that were listed in Activity 1 to start the discussion of conditions that influence the uncertainty of events. For example, conditions that affect the meteorologist's accuracy in predicting rain on a particular day may depend on past training, past experience, available information about present weather conditions in other areas, and unpredictable conditions related to pressure changes in the atmosphere.

2. Ask students to discuss the conditions that could influence whether the princess's lover will choose the door with the fair maiden or the door with the tiger. The conditions are likely to revolve around the students' perceptions of the princess's personality and that of her father. For example, the level of certainty may depend on such conditions as these:

How strongly does the princess's nature reflect the barbaric nature of her father?

How strongly does the princess want to save her lover's life?

Does the young man believe that the princess will guide him to the lifesaving door?

Activity 3: Polling the Students

1. Pose this question to the students:

Which door do you think the princess indicated to her lover?

Report the results as a whole class. Then report the results for boys and girls separately (e.g., 10/15, or 0.67, of the girls believe that the princess will point to the door with the lady). Make conjectures to explain any differing results.

Evaluation. Ask students to write a tightly organized argument making a case for their conjectures about why boys and girls would respond differently. Assess the papers on the basis of writing techniques (effective use of paragraphs, well-constructed sentences) and content (the logic of the argument). Seek help from the language arts teacher if you are not familiar with grading written papers for form and content.

COMMUNICATION: Integrating literature, writing, and mathematics provides opportunities for a focus on communication through the written medium. Another way to incorporate writing experiences into the middle school students' mathematics classes is to give them a graph or table from the newspaper and ask them to write the accompanying article. Compare and contrast their article with the original write-up from the paper.

2. Pose this question to the students:

What number between 0 and 1 would you use to represent your level of certainty that the young man will live? (1 means that life is certain, and 0 means that death is certain.)

Ask students to organize, represent, and analyze the resulting data using any statistical techniques that they have learned. Students can consider the class as a whole and then split the information for boys and girls. Groups of students can form conjectures and explanations based on the story, the students' perceptions of the story, the author's perceived intentions, and so on.

Activity 4: Making Predictions Given Certain Conditions

Ask students to consider different conditions in assigning varying levels of likelihood that the young man will live. (This is building the language associated with conditional probability.) The following are some classroom examples:

> "Let's suppose that Bridget is correct. Given (i.e., assuming) that the princess is more in love than she is jealous, how likely is it that the young man will live?"
>
> "Let's suppose that William is correct. Given (i.e., assuming) that the princess is more jealous than she is in love, what is the likelihood that the young man will live?"

Students should discuss whether they would expect overall levels of likelihood to go up or down on the basis of the different assumptions.

Activity 5: Designing a New Justice System

1. Propose to students that the king decided that his justice system should reflect the severity of the alleged crime. On one hand, murder should be dealt with more harshly than robbery. So, the king decided to use three doors in his arena for murder suspects. Behind two doors are hungry tigers, whereas behind one door is a lovely woman. For robbery, on the other hand, he kept the system as it was.

2. Assign students to work in groups on a revised system of justice where they list various crimes, rank them, and devise combinations of doors with tigers and ladies to reflect the different levels of severity. Each group would need to justify how their combinations reflect their ranking of the crimes.

3. Use the opportunity to state formally the probability of selecting a tiger in each situation. Then, students can write all the probabilities in decimal form and see whether the order matches their ranking of the crimes.

4. On the basis of making up new laws and punishments for crimes,

investigate how different governments create, interpret, and enforce laws.

Closing Comments

Although integrating literature and statistics may seem unusual, making these connections is natural and easy to do. Many literary examples exist in which some quantitative idea or probabilistic occurrence is a focal point for the story. These occasions can be used to help students see ideas about chance and data in varied contexts, building strength in their understandings. For further ideas to link statistics and probability with English studies, see Zawojewski, Nowakowski, and Boruch (1988), Boruch and Zawojewski (1987), and Smeeton and Smeeton (1985).

CONCLUDING COMMENTS

Exploring data and chance in school should be as natural as it is in everyday life. By building on the natural tendencies of students to notice and learn from information, we can implement the *Curriculum and Evaluation Standards*. We should begin in a world of experience and contexts that are relevant and familiar to our students. Students who formulate their own questions, gather and manipulate their own data, and analyze their information for decision making are being given an opportunity to experience the world through the eyes of a statistician. When experiences allow children to create their own methods of communication, find their own ways to solve problems, develop their own reasoning processes, and see how the quantitative processes connect to many other areas, children can construct their own world of mathematics. They gain confidence and power to succeed in a world that requires more and more people to be quantitatively literate. The *Curriculum and Evaluation Standards* affords an opportunity for improving and changing our methods of instruction as we strive to reach this goal.

♦ ♦ ♦ ♦ ♦ ♦ ♦ ♦

BIBLIOGRAPHY

AIMS K–4, K–6, and 5–9 Series. Fresno, Calif.: AIMS Educational Foundation, 1988–1989.

Artzt, Alice F., and Claire M. Newman. *How to Use Cooperative Learning in the Mathematics Class*. Reston, Va.: National Council of Teachers of Mathematics, 1990.

Baroody, Arthur J. "One Point of View: Manipulatives Don't Come with Guarantees." *Arithmetic Teacher* 37 (October 1989): 4–5.

Bezuk, Nadine. *Understanding Rational Numbers and Proportions*. Addenda Series, Grades 5–8. Reston, Va.: National Council of Teachers of Mathematics, forthcoming.

Boruch, Robert F., and Judith S. Zawojewski. "Coupling Literature and Statistics." *Teaching Statistics* 9 (May 1987): 34–36.

Burrill, Gail. "Implementing the Standards: Statistics and Probability." *Mathematics Teacher* 83 (February 1990): 113–18.

Cameron, Dwayne. "The American Statistical Prize Competition." *The Statistics Teacher Network* (January 1987): 1.

———. "Third Annual Statistical Prize Competition 1989 Results." *The Statistics Teacher Network* (February 1990): 1–2.

Charles, Randall, Frank Lester, and Phares O'Daffer. *How to Evaluate Progress in Problem Solving*. Reston, Va.: National Council of Teachers of Mathematics, 1987.

Choate, Stuart A. "Activities in Applying Probability Ideas." *Arithmetic Teacher* 26 (February 1979): 40–42.

Collis, Betty. "Simulation and the Microcomputer: An Approach to Teaching Probability." *Mathematics Teacher* 75 (October 1982): 584–87.

Corwin, Rebecca, and Susan Friel. *Used Numbers—Statistics: Prediction and Sampling, Grades 5–6*. Menlo Park, Calif.: Dale Seymour Publications, 1990.

Curcio, Frances R. *Developing Graph Comprehension: Elementary and Middle School Activities*. Reston, Va.: National Council of Teachers of Mathematics, 1989.

Davidson, Neil, ed. *Cooperative Learning in Mathematics*. Menlo Park, Calif.: Addison-Wesley Publishing Co., 1989.

Geddes, Dorothy. *Geometry in the Middle Grades*. Addenda Series, Grades 5–8. Reston, Va.: National Council of Teachers of Mathematics, forthcoming.

Gnanadesikan, Mrudulla, Richard L. Scheaffer, and Jim Swift. *The Art and Techniques of Simulation*. Palo Alto, Calif.: Dale Seymour Publications, 1987.

Goldberg, Howard, and Philip Wagreich. *Counting Out πr^2*. Chicago: Teaching Integrated Mathematics and Science (TIMS) Project, University of Illinois, 1989.

Hartoonian, H. Michael. "Social Mathematics." In *From Information to Decision Making: New Challenges for Effective Citizenship*. Edited by M. Laughlin et al., pp. 51–63. Washington, D.C.: The National Council for the Social Studies, 1989.

Hawkins, Anne S. "Student Surveys in the United Kingdom Annual Applied Statistics Competition." In *American Statistical Association 1987 Proceedings of the Section on Survey Research Methods*, pp. 372–74. Alexandria, Va.: The Association, 1987.

Hawkins, Anne S., and Ramesh Kapadia. "Children's Conceptions of Probability—A Psychological and Pedagogical Review." *Educational Studies in Mathematics* 15 (1984): 349–77.

Hinders, Duane C. "Monte Carlo, Probability, Algebra, and Pi." *Mathematics Teacher* 74 (May 1981): 335–39.

Hirsch, Christian R., and Glenda Lappan. "Transition to High School Mathematics." *Mathematics Teacher* 82 (November 1989): 614–18.

Holmes, Peter, N. Rubra, and D. Turner. *Statistics in Your World*. Berks, England: W. Foulsham & Co., 1980.

Huff, Darrell. *How to Lie with Statistics.* New York: W. W. Norton & Co., 1954.

Kahneman, Daniel, Paul Slovic, and Amos Tversky, eds. *Judgment under Uncertainty: Heuristics and Biases.* Cambridge: Cambridge University Press, 1982.

Kruskal, William. "Formulas, Numbers, Words: Statistics in Prose." *American Scholar* 47 (Spring 1978): 223–29.

Landwehr, James M., Jim Swift, and Ann E. Watkins. *Exploring Surveys and Information from Samples.* Palo Alto, Calif.: Dale Seymour Publications, 1987.

Landwehr, James M., and Ann E. Watkins. *Exploring Data.* Palo Alto, Calif.: Dale Seymour Publications, 1986.

Lappan, Glenda, Elizabeth Phillips, Mary Jean Winter, and William Fitzgerald. "Area Models for Probability." *Mathematics Teacher* 80 (March 1987): 217–23.

Lappan, Glenda, and Mary J. Winter. "Probability Simulation in Middle School." *Mathematics Teacher* 73 (September 1980): 446–49.

Lester, Frank K., and Diana Lambdin Kroll. "Implementing the *Standards*: Evaluation: A New Vision." *Mathematics Teacher* 84 (April 1991): 276–84.

Mathematics Resource Project. *Statistics and Information Organization.* Palo Alto, Calif.: Creative Publications, 1978.

McCord, David. "In the Middle." In *The Star in the Pail.* Boston, Mass.: Little, Brown & Co., 1975.

National Council of Teachers of Mathematics. *An Agenda for Action.* Reston, Va.: The Council, 1980.

———. *Curriculum and Evaluation Standards for School Mathematics.* Reston, Va.: The Council, 1989.

———. Curriculum and Evaluation Standards for School Mathematics Addenda Series, Grades K–6, edited by Miriam A. Leiva. Reston, Va.: The Council, forthcoming.

———. Curriculum and Evaluation Standards for School Mathematics Addenda Series, Grades 5–8, edited by Frances R. Curcio. Reston, Va.: The Council, forthcoming.

———. Curriculum and Evaluation Standards for School Mathematics Addenda Series, Grades 9–12, edited by Christian R. Hirsch. Reston, Va.: The Council, forthcoming.

———. *Mathematics Teacher* 83 (February 1990). Focus issue: Data Analysis.

———. *Professional Standards for Teaching Mathematics.* Reston, Va.: The Council, 1991.

———. *Teaching Statistics and Probability,* 1981 Yearbook of the National Council of Teachers of Mathematics, edited by Albert P. Shulte. Reston, Va.: The Council, 1981.

Newman, Claire M., Thomas E. Obremski, and Richard L. Scheaffer. *Exploring Probability.* Palo Alto, Calif.: Dale Seymour Publications, 1987.

Nisbett, Richard E., and Lee Ross. *Human Inference: Strategies and Shortcomings of Social Judgment.* Englewood Cliffs, N. J.: Prentice-Hall, 1980.

Paulos, John Allen. *Innumeracy.* New York: Hill and Wang, 1989.

Phillips, Elizabeth. *Patterns and Functions.* Addenda Series, Grades 5–8. Reston, Va.: National Council of Teachers of Mathematics, 1991.

Phillips, Elizabeth, Glenda Lappan, Mary Jean Winter, and William Fitzgerald. *Middle Grades Mathematics Project: Probability.* Menlo Park, Calif.: Addison-Wesley Publishing Co., 1986.

Reys, Barbara. *Developing Number Sense in the Middle Grades.* Addenda Series, Grades 5–8. Reston, Va.: National Council of Teachers of Mathematics, 1991.

Rowan, Thomas, (Chair). Report on the Task Force of Addenda to the NCTM K–12 Curriculum and Evaluation Standards for School Mathematics. Unpublished report, November 1988.

◆ ◆ ◆ ◆ ◆ ◆ ◆ ◆

Russell, Susan Jo, and Rebecca B. Corwin. *Used Numbers—the Shape of the Data, Grades 4–6.* Palo Alto, Calif.: Dale Seymour, 1989.

Simon, Julian, and Allen Holmes. "A New Way to Teach Probability and Statistics." *Mathematics Teacher* 62 (April 1969): 283–88.

Smeeton, N. C., and Eunice A. Smeeton. "Statistical Ideas in English Studies." *Statistics Teacher* 7 (November 1985): 34–37.

Statistics Teacher Network. Newsletter published by the American Statistical Association-National Council of Teachers of Mathematics Joint Committee on the Curriculum in Statistics and Probability. (Rose-Hulman Institute of Technology, 5500 Wabash Avenue, Terre Haute, IN 47803)

Stats: The Magazine for Students of Statistics. Washington, D.C.: American Statistical Association.

Steen, Lynn Arthur. "Teaching Mathematics for Tomorrow's World." *Educational Leadership* 47 (September 1989): 18–22.

Stockton, Frank. "The Lady or the Tiger." In *Adventures in Reading*, pp. 19–24. Orlando, Fla.: Harcourt Brace Jovanovich, 1980.

Teaching Statistics. Longman Group UK Limited, England.

Travers, Kenneth J. "Using Monte Carlo Methods to Teach Probability." In *Teaching Statistics and Probability*, 1981 Yearbook of the National Council of Teachers of Mathematics, edited by Albert P. Shulte, pp. 210–19. Reston, Va.: The Council, 1981.

Travers, Kenneth J., and Kenneth G. Gray. "The Monte Carlo Method: A Fresh Approach to Teaching Probabalistic Concepts." *Mathematics Teacher* 74 (May 1981): 327–34.

Tufte, Edward R. *The Visual Display of Quantitative Information.* Cheshire, Conn.: Graphics Press, 1983.

U.S. Bureau of the Census. *1990 Census Education Project.* Washington D.C.: U.S. Bureau of the Census, 1990.

Viorst, Judith. "Who's Who." In *If I Were in Charge of the World.* New York: Atheneum Publishers, 1981.

Vygotsky, Lev. *Thought and Language.* Cambridge, Mass.: MIT Press, 1934/1986.

Watkins, Ann E. "Monte Carlo Simulation: Probability the Easy Way." In *Teaching Statistics and Probability,* 1981 Yearbook of the National Council of Teachers of Mathematics, edited by Albert P. Shulte, pp. 203–9. Reston, Va.: The Council, 1981.

Zawojewski, Judith S., Jeri Nowakowski, and Robert F. Boruch. "Romeo and Juliet: Fate, Chance, or Choice?: An English Lesson Using Probability." *Teaching Statistics* 10 (May 1988): 37–42.

Computer Software

A Chance Look. Sunburst Communications, 39 Washington Ave., Pleasantville, NY 10570-2898.

This software simulates events for standard or student-designed cubes and spinners having equally likely or weighted outcomes. Outcomes from combined events and sums of cubes can be displayed. Counts, a tally chart, and a list of outcomes can be viewed as the experiment is carried out. Final results can be displayed in relative frequency charts, bar graphs, and line graphs.

Appleworks. Apple Computer, 20525 Mariani Ave., Cupertino, CA 95014.

This software contains both spreadsheet and data base programs that are very easy to use. Data of any type are easily entered into the spreadsheet cells, and many useful formulas are built into the program. The data base has a large number of options for entering, displaying, organizing, and selecting data.

Bank Street Filer. Scholastic Software, 730 Broadway, New York, NY 10003.

This data base software is menu driven, so users do not have to remember the command keys needed to manipulate, search, or sort the data.

Data Insights. Sunburst Communications, 39 Washington Ave., Pleasantville, NY 10570-2898.

This graphing software displays data in a variety of forms, including box plots, stem-and-leaf plots, and scatter plots. Data entry is simple, and conversions between graphs are easy. The software has been developed for the Apple II family, IBM PC, IBM PS/2, and Tandy 1000.

Easy Graph. Houghton Mifflin Publishing Co., One Beacon St., Boston, MA 02108.

This graphing program was developed for the elementary school level and allows the user to develop pictographs, bar graphs, and pie graphs. Graphs can be converted from one type of display to another. The program is easy to use, but limits the amount of data that can be entered and the number of bars that can be displayed.

MECC Create-a-Base. MECC, 3490 Lexington Ave. North, Saint Paul, MN 55112.

This data base software is a step-by-step menu-driven program with large text and an easy-to-read screen format. It has many standard features and is easy to learn and use.

PFS: Graph. Software Publishing Corporation, 1901 Landings Dr., Mountain View, CA 94043.

This graphing package allows the simultaneous display of up to four bar and line graphs and the creation and display of pie graphs. Data entry is easy, and conversion between graphic modes is a simple one-step process.

Surveys Unlimited. Learning Well, 200 South Service Rd., Roslyn Heights, NY 11577.

This software program allows students to gather data on several prepared surveys, make their own surveys, see the survey results in table and graph form, and print out surveys and results. The questions can be answered in multiple choice, numerical, or word form, and the graphing capabilities include bar, line, and circle graphs.

Quantitative Literacy Series Software. Palo Alto, Calif.: Dale Seymour Publications, 1989.

This software package runs on the Apple II family. (There are plans to make it available to IBM PC users.) It provides visual graphics capabilities and simulation models to go along with the four books in the QL series.

BLACKLINE MASTER 1
IS THIS GAME FAIR?

Is This Game Fair?

A game for two participants. Each team will need a pair of number cubes.

RULES:
1. One of the participants is the PLAYER and the other is the OPPONENT. Only the PLAYER rolls the number cubes.
2. Each participant starts out with ten points.
3. Each time the PLAYER rolls a sum of 7, the OPPONENT must give up or transfer three points to the PLAYER.
4. Each time the PLAYER rolls any sum other than 7, he or she must give up one point to the OPPONENT.
5. Record the results of each roll on this game sheet. Record how many points the PLAYER or the OPPONENT has at the end of each roll of the dice.
6. The student with the most points at the end of ten rolls is the winner. If one of the participants runs out of points before ten rolls, the other participant is the winner.

BEFORE YOU BEGIN:
IS THIS GAME FAIR?
WHY?

	Start	Roll 1	Roll 2	Roll 3	Roll 4	Roll 5	Roll 6	Roll 7	Roll 8	Roll 9	Roll 10
	Sum of 7?	Yes or No	Yes or No	Yes or No	Yes or No	Yes or No	Yes or No	Yes or No	Yes or No	Yes or No	Yes or No
PLAYER	10										
OPPONENT	10										

BLACKLINE MASTER 2
WHO WON?

WHO WON?
LET'S RECORD OUR RESULTS

	Team 1	Team 2	Team 3	Team 4	Team 5	Team 6	Team 7	Team 8	Team 9
PLAYER									
OPPONENT									

	Team 10	Team 11	Team 12	Team 13	Team 14	Team 15	Team 16	Team 17	Team 18
PLAYER									
OPPONENT									

BLACKLINE MASTER 3
STUDENT SURVEY ON FUTURE PLANS

Directions: For each question, darken the circle next to one choice. Your answers will be completely confidential; only summary data will be reported. THANK YOU for taking the time to complete this survey. Getting your answers and those from others is important to produce accurate information on this important topic.

1. How old are you?

- ○ Less than 12
- ○ 12
- ○ 13
- ○ 14
- ○ 15
- ○ 16
- ○ 17
- ○ 18
- ○ 19
- ○ Older than 19

2. What is your sex?

- ○ Male
- ○ Female

3. Do you plan to get married?

- ○ Yes
- ○ No

4. Do you plan to have children?

- ○ Yes
- ○ No (If no, skip to Question 6.)

5. How many children would you like to have?

- ○ 1
- ○ 2
- ○ 3
- ○ 4
- ○ 5
- ○ 6
- ○ 7 or more

6. After high school, which of the following do you plan to do?

- ○ Attend a 2-year college
- ○ Attend a 4-year college
- ○ Go to a trade or vocational school
- ○ Join the Armed Forces
- ○ Get a full-time job
- ○ None of these

7. Of the following occupations, which *one* would you *most* like to pursue after school?

- ○ Doctor
- ○ Teacher
- ○ Social Worker
- ○ Lawyer
- ○ Computer Programmer
- ○ Stock Broker
- ○ Fire Fighter
- ○ Hairdresser
- ○ Mechanic
- ○ Carpenter
- ○ Truck Driver
- ○ Farmer
- ○ Forest Ranger
- ○ None of these

8. Of the following occupations, which *one* would you *least* like to pursue after school?

- ○ Doctor
- ○ Teacher
- ○ Social Worker
- ○ Lawyer
- ○ Computer Programmer
- ○ Stock Broker
- ○ Fire Fighter
- ○ Hairdresser
- ○ Mechanic
- ○ Carpenter
- ○ Truck Driver
- ○ Farmer
- ○ Forest Ranger
- ○ None of these

BLACKLINE MASTER 4

These items are the first draft of a survey prepared by a seventh-grade class. Read each question carefully. Rewrite each item as you think it should appear on the final draft. Explain any changes you make.

SURVEY ON CALCULATORS IN THE JUNIOR HIGH SCHOOL

1. How old are you?____________

2. Do you consider yourself a good mathematician? yes no

3. Which type of calculator do you prefer to use?
   a. solar powered
   b. battery powered
   c. scientific
   d. other:___________________

4. How often do you use a calculator and for what purposes?

   ____

   ____

   ____

5. Based on recent research, it has been found that technological advancement in today's society will eventually affect the pedagogical ideology relating to the use of electronic devices in schools. Do you agree or disagree?

   agree disagree

6. No student who has not demonstrated success in basic skills should be permitted to use a calculator in the classroom. Do you agree or disagree?

   agree disagree

7. Should immature junior high school students be exposed to the mindless activity of using a calculator to solve mathematical problems?

   yes no

8. When should calculators be used in school?

   ____

   ____

   ____

BLACKLINE MASTER 5
COMPARING VISUAL REPRESENTATIONS

The following displays are based on the same set of data, in which twenty-five students first estimated and then counted the number of raisins in 1/2-ounce boxes.

Back-to-Back Stem-and-Leaf Plots

Actual Ones	Tens	Estimates Ones
	1	6
9	2	0000223555
9999888855422221	3	00015
11100000	4	00005
	5	0000

9 | 2 | 0 means 29 actual count
20 estimate

Double Bar Graph (NOTE: Incomplete data set shown.)

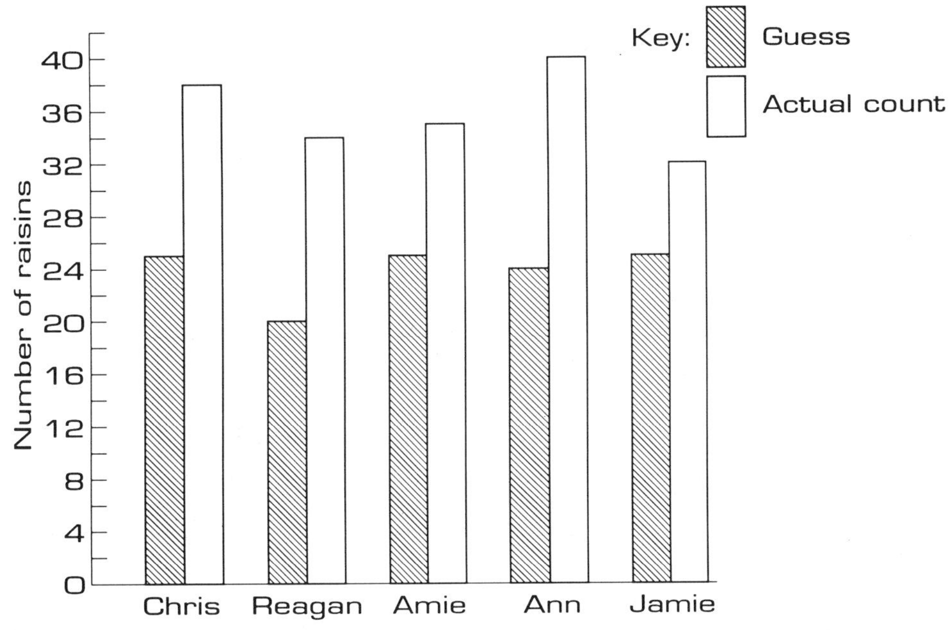

Multiple Box Plots

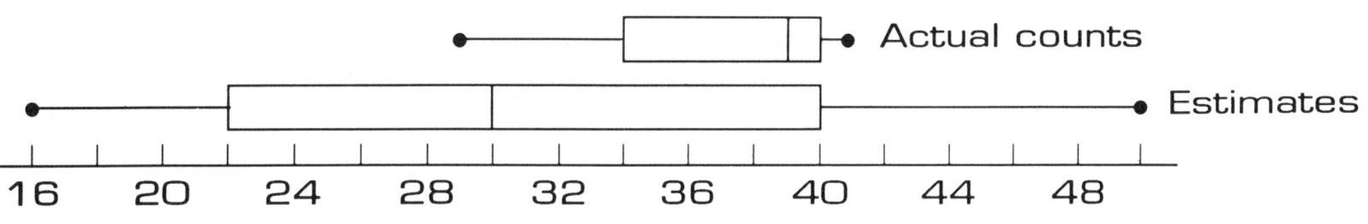

BLACKLINE MASTER 6
A LOOK AT THE AVERAGE WAGE

The head of the employees' union in the Millstone Manufacturing and Marketing Company was negotiating with Mr. Millstone, the president of the company. He said, "The cost of living is going up. Our workers need more money. No one in our union earns more than $18 000 a year."

Mr. Millstone replied, "It's true that costs are going up. It's the same for us—we have to pay higher prices for raw materials, so we get lower profits. Besides, the average salary in our company is over $22 000. I don't see how we can afford a wage increase at this time."

That night the union official conducted the monthly union meeting. A sales clerk spoke up. "We sales clerks make only $10 000 a year. Most workers in this union make $15 000. We want our pay increased to at least that level."

The union official decided to take a careful look at the salary information. He went to the payroll department. They told him that they had all the salary information on a spreadsheet in the computer, and they printed off this table:

Type of job	Number employed	Salary	Union member
President	1	$250 000	No
Vice-president	2	130 000	No
Plant Manager	3	55 000	No
Foreman	12	18 000	Yes
Workman	30	15 000	Yes
Payroll Clerk	3	13 500	Yes
Secretary	6	12 000	Yes
Sales Clerk	10	10 000	Yes
Custodian	5	8 000	Yes
TOTAL	72	$1 593 500	—

The union official calculated the mean:

$$\text{MEAN} = \frac{\$1\ 593\ 500}{72} \approx \$22\ 131.94$$

BLACKLINE MASTER 7

“Hmmm,” he thought, “Mr. Millstone is right, but the mean salary is pulled up by those high executive salaries. It doesn’t give a really good picture of the typical worker’s salary.”

Then he thought, “The sales clerk is sort of right. Each of the thirty workmen makes $15 000. That’s the *most common* salary—the mode. However, there are thirty-six union members who don’t make $15 000—of those, twenty-four make less.”

Finally, the union official said to himself, “I wonder what the *middle* salary is?” He thought of the employees as being lined up in order of salary, low to high. The middle salary (it’s called the *median*) is midway between employee 36 and employee 37—36 below and 36 above. He said, “Employee 36 and employee 37 each make $15 000, so the middle salary is also $15 000.”

Questions

1. If the twenty-four lowest salaried workers were all moved up to $15 000, what would be

   a. the new median?
   b. the new mean?
   c. the new mode?

2. What salary position do you support, and why?

BLACKLINE MASTER 8
EXPLORING STANDARD DEVIATION

Suppose that light bulbs from three companies were randomly selected and tested for longevity. Testers recorded the number of hours that the bulbs burned and made the plots below. Examine, compare, and contrast the three sets of data listed below. Which statistical descriptive terms can be found by inspection (with little or no calculation) of the numbers? For the term(s) that can be determined, record the value(s) in the table below. For the term(s) that cannot be determined easily, estimate the value. Use two different types of writing instruments (such as a pen and a pencil) so that you can tell which ones you estimated.

Company A

```
2 | 7 9
3 | 2 3 9
4 | 5 6
5 | 7 8 9
6 | 1 5 8 9
7 | 3 4 7 9
8 | 3 5 6 7 8 9 9
9 | 2 6 7 8 9          (2 | 7 means 27)
```

Company B

```
5 | 4 7 8
6 | 0 0 0 2 2 4 5 5 5 6 8 8 9
7 | 1 1 2 3 4 5 5 5 6 6 7
8 | 2 9
9 | 1
```

Company C

```
3 | 7
4 | 5
5 | 1 5 8 9 9
6 | 0 1 2 2 3 3 5 5 8
7 | 0 0 2 2
8 | 0 3 4 4 5 6 7 9
9 | 2 3
```

VALUES FROM **INSPECTION** AND **ESTIMATION**

	No. of tests	Range	Mean	Median
Company A				
Company B				
Company C				

BLACKLINE MASTER 9

Using a scientific calculator in the statistics mode, enter the data for Company A. Find and record in the following table those values that you estimated. How close were you? Now find and use the key that gives you the standard deviation. Record that value. Repeat this procedure for Companies B and C.

VALUES FROM **INSPECTION** AND **CALCULATION**

	No. of tests	Range	Mean	Median	Standard deviation
Company A					
Company B					
Company C					

Discuss similarities and differences among the sets of data and the statistical values. Go back to the original numbers to see if you can decide what standard deviation might mean.

The calculator found the standard deviation using a formula that involves subtracting each number from the mean, squaring each of those differences, adding all of them together, dividing by the number of tests, and then finding the square root. But rather than do all that work, we use the calculator and focus on the meaning of standard deviation.

We need to subtract the standard deviation from the mean and then add the standard deviation to the mean. Record. For example, if a set of data had a mean of 100 and a standard deviation of 10, our calculations would give us 90 and 110.

In normally distributed data, 68 percent (approximately 2/3) of the numbers will be between these two numbers. Go back to the chart and find the interval that should contain 2/3 of the numbers. Test to see if it does.

BLACKLINE MASTER 10
COINCIDENCE, SUPERSTITION, OR FACT?

If You Understand Pizza, You Understand Subway Fares

By George Fasel

In 1975, when I first visited New York City, the transit fare was 35 cents for service that, in my experience, ran from indifferent to poor. By 1978, when I returned, the transit fare had increased to 50 cents. It also seemed to me that service had deteriorated.

It was explained to me, however, that this was my fault: Since I had been away for three years, the inescapable decline in the service was bound to seem dramatic. If I had experienced that decline on a daily basis, I would scarcely have noticed it.

In 1979, I moved to New York. I had no choice but to become a regular subway rider, and I took comfort in the knowledge that I would barely register the system's erosion. The next year, however, there was a transit strike, which was not my fault. Unavoidably, exposure to an alternate means of transportation dramatized the decay of the subway. I walked for 11 days, and, while the experience was inconvenient, on balance I was happier.

A lesson in the most elementary economics

When the fare reached 90 cents in 1984, I was pretty steamed. After all, by the Metropolitan Transportation Authority's own evaluations, service was getting worse. Breakdowns, delays, accidents, crime and unspeakable indignities now came with the fare at no extra charge. If riders in rush hour had been treated instead like cattle being led to slaughter, they would have wondered how to account for the improvement in their status.

The experience was bad enough, but the principle was worse. The product was going to pot, and I was supposed to pay more for it.

A friend came to the rescue. He is a man who understands the delicate interplay of markets. "You've missed the connection," he told me. "The transit token has no relationship to capital costs, union contracts, passenger miles, or depreciation schedules. Forget all that. The critical variables are flour, tomato sauce and mozzarella cheese."

"You mean . . . ?"

"Yes, the token is subtly but inextricably linked to the price of a slice of pizza. Don't ask why. It just is. Where, today, in 1984, can you get a decent slice of pizza in New York for 75 cents? It's tough to swallow"—I think he meant the truth, not the pizza—"but that's how this town works."

Not long ago, the press was full of stories alleging epic waste and bungling in the M.T.A. (the 63d Street line, missed Federal funding opportunities, the usual sort of thing). Perversely, such stories seem to reach a peak just before we poor suckers who pay the fare get stuck with a new increase. I stiffened, determined to resist, or at least to protest. But then late one evening I stopped for a slice of pizza at the parlor across the street from my apartment. For nearly two years, I had been accustomed to laying down my dollar bill and picking up my slice and a dime. This time, I received no change.

There it is, fellow passengers. Oh, you can fight it if you wish, and I wouldn't blame you, but you're bucking an iron law, apparently as inflexible as the law of gravity is in its sphere. Mario Cuomo has decided it is futile even to attempt to save the 90-cent fare, because the Governor is a man who knows the price of a slice.

If you don't believe this fare-pizza thesis, ask yourself, on Jan. 1, when the price of a subway and bus ride becomes $1, what other explanation there is for the increase.

George Fasel is a vice-president of the Bankers Trust Company.

Dec. 14, 1985 (Op-Ed) p. 27

BLACKLINE MASTER 11
COINCIDENCE, SUPERSTITION, OR FACT?

Bad news for cola-quaffing athletes

Dr. Grace Wyshal, of Harvard University's School of Public Health, has found indications that a lifetime of drinking carbonated beverages may leave athletic women with weaker bones than nonathletic women and those who do not drink colas.

"There seems to be some interaction between carbonated drinks and exercise," she says, explaining the results of a study that looked at the daily habits of 5,398 women.

Normally, exercise helps women build strong bones, but the college athletes who drank soft drinks were about twice as likely as those who didn't to suffer a first bone fracture after the age of 40, when women's bones start to become more brittle.

What's worrisome in colas is not the caffeine they contain, but phosphoric acid, which may interfere with the absorption of bone-building calcium, Wyshal says.

Or it simply may be that the women are not drinking milk, Wyshal says, noting that American milk consumption has dropped steadily over the last 20 years.

"I think it's important to know that younger people drink far more [carbonated beverages] than older people," she says.

"It ranged from about 23 ounces a day among the former college athletes under 30 to about 16 ounces a day for those 60 years and older."

Dear Abby: My friends and I have a question that only you can answer. When you catch a bouquet at a wedding and the marriage ends in divorce, are you still going to be the next to be wed?

Just Wondering

Dear Wondering: The catcher has no guarantee that she will be the next to wed regardless of how the marriage turns out.